AF321685

Do Less Harm

GLOBAL STUDIES IN MEDICINE, SCIENCE, RACE, AND COLONIALISM

Ahmed Ragab, *Series Editor*

Do Less Harm

ETHICAL QUESTIONS FOR HEALTH HISTORIANS

Edited by
Courtney E. Thompson
and Kylie M. Smith

Johns Hopkins University Press
Baltimore

© 2025 Johns Hopkins University Press
All rights reserved. Published 2025
Printed in the United States of America on acid-free paper

2   4   6   8   9   7   5   3   1

Johns Hopkins University Press
2715 North Charles Street
Baltimore, Maryland 21218
www.press.jhu.edu

Library of Congress Cataloging-in-Publication Data

Names: Thompson, Courtney E., editor. | Smith, Kylie, 1968– editor.
Title: Do less harm : ethical questions for health historians /
edited by Courtney E. Thompson, Kylie M. Smith.
Description: Baltimore : Johns Hopkins University Press, 2025. |
Series: Global studies in medicine, science, race, and colonialism |
Includes bibliographical references and index.
Identifiers: LCCN 2024055574 (print) | LCCN 2024055575 (ebook) |
ISBN 9781421452265 (paperback ; alk. paper) |
ISBN 9781421452272 (ebook)
Subjects: MESH: History of Medicine | Historiography |
Ethics | Writing | Teaching—ethics
Classification: LCC R133 (print) | LCC R133 (ebook) |
NLM WZ 40 | DDC 610.9—dc23/eng/20250307
LC record available at https://lccn.loc.gov/2024055574
LC ebook record available at https://lccn.loc.gov/2024055575

A catalog record for this book is available from the British Library.

*Special discounts are available for bulk purchases of this book. For more
information, please contact Special Sales at specialsales@jh.edu.*

EU GPSR Authorized Representative
LOGOS EUROPE, 9 rue Nicolas Poussin, 17000, La Rochelle, France
E-mail: Contact@logoseurope.eu

*For our historical actors, students, publics, colleagues,
and descendant communities, past, present, and future,
to whom we pledge to do less harm*

CONTENTS

PART V TEACHING

ACKNOWLEDGMENTS

As editors of such a large volume, we have many people to thank, not least all of our colleagues who worked so hard to bring this together. We are incredibly grateful for the support of the Consortium for History of Science, Technology and Medicine, which helped with the logistics for the working group over two years and without which this community of scholars would not exist. We also want to thank Matt McAdam at Johns Hopkins University Press for his support of our big idea and particularly to thank Ahmed Ragab, who saw the potential immediately, advocated for a longer word count, and helped wrangle issues along the way.

I (Kylie) would like to thank the women in our online writing group for their support, community, and constant cheer squad, and of course Cookie for the welcome distraction. I owe a debt of gratitude to Courtney for inviting me on this journey and being such a well-organized coeditor and to all of the interlocutors in the working group and at the American Association for the History of Medicine who have shaped my thinking so profoundly. I also would like to thank my intellectual home without which this work would not have been possible: Emory University in Atlanta, which is built on the unceded territory of the Muscogee Creek people.

I (Courtney) would likewise like to thank the women of our online writing group; Kylie, my intrepid coeditor, for her patience and big-picture thinking; my dogs, Winnie and Gigi, both of whom joined the family during the process of putting together first this working group and then this volume; my mother, Sharon Thompson; and Scott DiGiulio, for listening to countless monologues about our working group discussions and this volume. I worked on this volume at Mississippi State University, which is on the unceded land of the Choctaw and Chickasaw peoples.

Do Less Harm

Introduction

A Field Guide to Doing Less Harm

COURTNEY E. THOMPSON AND KYLIE M. SMITH

> Ethical is a much more expansive word than moral, because ethical allows for the mistakes people make.
>
> *—bell hooks*[1]

It was May 2021. One year into the COVID-19 pandemic, and the global death toll had just passed three million. It was one year after the explosion of Black Lives Matter (BLM) protests in the wake of the murder of George Floyd and several months into Russia's invasion of Ukraine. The American Association for the History of Medicine (AAHM), after canceling the 2020 meeting, scheduled a virtual meeting for 2021. We (the editors) had published our first books during the pandemic—Kylie's *Talking Therapy* in 2020 and Courtney's *An Organ of Murder* in 2021—so we both eagerly signed up to participate in the "author chats" organized during the meeting.

And then an interesting thing happened. In both of our author chat sessions, the conversation turned to matters of ethics—of how and why we do health history and what our ethical responsibilities are as health historians. We also belonged to the same writing group at the time, which might have informed the convergence of our thoughts; surely the pandemic experience and BLM movement were inflecting our collective discussions. Post-AAHM, a group of those present at the virtual meeting were invigorated by these discussions; further interest was generated via email and through social media. We quickly realized that there was a perfect opportunity to continue the conversation: the Consortium for History of Science, Technology and Medicine had recently issued a call for new working group proposals. We submitted a proposal, and the Ethics and/in the History of Medicine and the Human Sciences Working Group had its first meeting on August 12, 2021, just three months after AAHM.

We were the co-conveners of the working group, but from the beginning it was designed as a collective enterprise. As a group, we developed a plan for readings and discussions, keeping track of the readings and of future suggestions via shared Dropbox folders and Google Docs. One topic, and our tendency to disagree with one another about it, would lead to conversations that drifted elsewhere, suggesting future topics and readings. Our membership was similarly varied, with participants ranging from grad students to emeriti, scholars from across the globe (which made scheduling a challenge), and those with different training, research interests, and professional roles. This diversity of perspectives was essential to our developing community, leading to engagement with scholars and with ideas far outside health history.

Our working group was truly a *working* group; from the beginning, we collectively had some sense of wanting to *do* something with our work. Perhaps it would be a syllabus or a set of guidelines that we could ask the AAHM to endorse. But within the first few months our purpose became clear: we wanted to write a book. In spring 2022, the working group collectively brainstormed what topics and questions we would like to see and how we thought the book should be organized, which resulted in the framework we have used for this volume (more on that shortly). In an attempt to distribute topics fairly, we solicited preferences and then invited working group participants to write on one of their preferred topics. Then we began to open our circle by issuing invitations to other scholars who we knew could speak to some of the topics that had emerged as we planned the volume. By the end of the second year of the working group, our circle of scholars had expanded and drafts of chapters had been written, shared, and workshopped (twice).

Do Less Harm is the result of two years of conversations, collaborations, and labor. While Kylie and Courtney served as its editors, the design of this volume, its topics and themes, and its central ethos are features that emerged out of extensive collaboration. This book is as much a material artifact of the community that we built and the conversations we had as it is a collection of essays. These essays, while short and often informal or experimental, are representative of deep thought about and concern with the ethics of health history.

***

What do we mean by "ethics" in the context of the history of health and medicine? This is one of the questions that drove the working group, and our varied attempts to answer it (about which we rarely reached a consensus) are demonstrated in the breadth and variety of the chapters in the volume. The application of ethical practices, for our contributors, took different forms depending on the context of their own work and the nature of their experiences, but it was clear that none of us had had any formal training in ethics for historians, and that nothing in our education as historians had prepared us to think about ethics theoretically. Collectively, we began trying to define it through what we thought it was not.

We knew that we did not mean ethics as defined solely by the online training modules for applications to an institutional review board (IRB). For many of us, applying to an IRB or university ethics committee had been a frustrating exercise because the bias toward STEM (science, technology, engineering, and medicine) and health disciplines meant that we were often told we did not need an ethics ruling because what we were doing was not "generalizable" and therefore not even "research." This kind of encounter demonstrates how narrowly defined "ethics" has become when it is considered important only in relation to economic conflicts of interest or informed consent in human subjects research.[2] This approach left all of us cold, because it dismissed the importance of historical research as an enterprise, and because it assumed there was no harm that could be done by it. Of course, many of us are familiar with principles of bioethics and have worked on projects using more traditional qualitative methods of health science work, especially those approaches that seek to put lived experience at the center of health research.[3] There are some valuable lessons here for health historians, but these approaches ultimately fail to get to the root of our issues, which are related specifically to questions of historical method. Our contention in the working group and throughout this volume is that without incorporating ethical thinking in their work, historians can do harm when they fail to consider things like their own positionality, the politics of the archive, or the rights of historical subjects who were patients.

It was this last point that was our starting point for defining a practice of ethics. One of the first books we read together was Susan Lawrence's *Privacy and the Past,* which deals with the issues arising for historians from new patient privacy laws in the United States.[4] Lawrence's book was provocative for us because she argues that historians have the responsibility to push back against archival closures in the wake of HIPAA (the Health Insurance Portability and Accountability Act). This idea raised issues for our working group: *Is* that our responsibility?[5] We did not all agree, and the essays in this volume reflect our different answers to this question. But we did wonder if our pushing for access to records causes more or less harm, and to whom. We had many discussions about how being overly concerned with privacy can actually work to reinforce stigma. Who is harmed by the revelation of mental illness in a family two generations (or two centuries) ago, exactly? Does our squeamishness about that history serve to reinforce shame and stigma? Are any concerns about privacy covered by using fake names or anonymizing records? (See Imada's essay in this volume for more on this issue.) There are so many question marks in this paragraph because we did not have any clear answer to these complicated questions. Ultimately, our inability to answer them categorically is why you are holding this book in your hands.

We knew too that we did not just mean ethics in terms of social history. We worked our way through both Susan Reverby and David Rosner's original essay "Beyond 'the Great Doctors'" from 1979 and their revisitation of it published in 2004.[6] These pieces are foundational to the possibility of an ethical approach to the history of health care, because in the very formulation of "history from below," we are reminded to check our privilege and to consider how the health care encounter is lived, experienced, and embodied. Informed by E. P. Thompson and taken up by Roy Porter, this approach reminds us that the meanings of health and medicine are made by the people who experience them, by the relationships that are forged at the intersection of person, community, and practice.[7] But even in works as foundational as these, there were no real answers to the ethical, as opposed to methodological, issues we found ourselves facing. In some ways, history from below, or social history, was the cause of a whole new set of ethical issues because it encouraged us to look for patient or subaltern perspectives as though this was now an essential part of "good" history, without necessarily thinking through the impact of that quest on the people we were writing about. Social history was a necessary first step toward a more ethical practice, but it would not be our last. Indeed, as Ludmilla Jordanova has observed, "since historical knowledge is produced through social processes within communities and in specific contexts . . . those who undertake such work have ethical obligations as well as intellectual ones."[8] Thus, *all* historians engaging *all* historical methodologies have ethical obligations.

One of the immediate problems posed by the move toward social history was the problem of the archive itself. Across the working group, we soon saw that we all struggled with issues in the making of health and medical history that were the result of decades of unquestioned approaches and a sense of entitlement to the archives.[9] That entitlement also came with a kind of arrogance regarding archival veracity and the question of "truth," which took us deep into readings on power and silence in the archives. Reading against and across the grain, à la Ann Laura Stoler, was one approach we discussed, but we were also, and perhaps more deeply, informed by Michel-Rolph Trouillot's argument about the way that those archives were created in the first place, whose stories they considered worth telling, and how those kinds of decisions framed the very possibility of the archive for the subaltern.[10]

A thread emerged in our group discussions (and thus in the essays in this volume), which was the concept of "justice." We were not particularly self-aware about this, but it appeared again and again in different contexts and in response to various readings.[11] We knew we were annoyed by James Sweet's pronouncements about the so-called ills of presentism, and we found ourselves talking through how an ethical approach to the history of health care was in many ways a presentist one—and that we were fine with that.[12] In fact, we were animated by the idea that the history of health and medicine is important today precisely because it is able, in fact it *needs*, to inform discussions about the greatest problems at present in our health care systems—namely, health disparities, racism, ableism, and discrimination.

Given that we were an online working group formed during the height of the COVID-19 pandemic, our everyday lived experiences as people became inextricably entwined with our concerns as historians, researchers, archivists, and teachers. We make no apology for that, so this idea of justice—or the pursuit of it—has become an ethical imperative for many of the authors of this volume. We did not try to come to a consensus about what *justice* meant; indeed the chapters in this volume show many different interpretations of that term. But we also do not mean that histories that work backward from current social problems are the only ethical ones—far from it. Subject matter does not in and of itself constitute an ethical approach. Rather, our concern with justice was the use to which history could be put in the current moment and the ways that a concern with justice across the board shaped our methods and approaches. It is these methods and approaches that the authors in this volume have sought to address.

As historians, we were well aware of the existence of racism and disparities long before COVID-19 (and BLM) apparently revealed them to people who had not been paying attention, but we were also aware of how global inequities were playing out in ways that reflected the history of medicine itself. A growing body of work in global histories of medicine and health care demonstrates the ways that empire and colonialism laid the foundations for the fractures exposed by COVID-19.[13] Yet too many histories still fail to account for the ways that medicine and health care were not simply exported from the "metropole" to the "periphery." We grumbled about histories that failed to consider how medical and health knowledge was made at the intersection of global and colonial relations, negotiated by intermediaries, or resisted outright. We were annoyed by the way that some historians still tended to center both the perspective and the archives of imperial power, replicating the great doctor or nurse narrative. We turned to decolonial theory to try to shift our frame of reference and to think about the ways that a radical reorientation of perspective away from state actors and their official archives might provide some ethical insights.[14]

In many ways, this call to "decolonize" our methods is what spurred our concern with ethics in the first place and is why we think ethics is so important for historians of medicine. Scholars of and from the Global South and Indigenous communities raised questions about ethical research practices long before we thought of them, but somehow history as a discipline has largely held itself at remove from these debates. None of our training as historians of health or medicine had prepared us to deal with the impact of HIPAA, for example, and the removal and restriction of patient records from archives has made many of us rethink both our practices and our assumptions. What right do we actually have as historians to tell stories about the health of people in the past, and how do we balance our need to expose past harms with the needs of the people and communities we are writing about? The ripple effect of patient privacy legislation has made us think about what harm we might also be doing as historians and how we have been ignorant (willfully or otherwise) about the ethical implications of what we have always assumed to be our right.

While many of us have experienced the dubious joy of submitting an ethics committee or IRB proposal only to be told we were not actually doing research and were therefore exempt (see Ramos's essay for one such story), we were also aware that this kind of ruling fails to account for the way that historical knowledge is in fact generalizable and powerful. We can do harm as historians of health and medicine, precisely because questions of health and medicine are necessarily so personal. They deal with other people's feelings, reactions, perceptions, and bodies, not simply "the facts"—facts, after all, are themselves socially constructed. If social history is what pushed us to think beyond the great doctors, then it is disability studies that has pushed us to think beyond the great historian narrative and to question our own credentials and approaches.

Scholars studying the history of disability have long wrestled with their own questions of ethical practices born of the call for "nothing about us without us" by disability activists. While initially a call for advocacy and representation in policy development, this idea has extended to the scholars of the field. Debates within disability studies have examined issues of positionality, anonymity, and racism and have raised important ethical questions about how research and representation are done.[15] Because health, medical, nursing, and psychiatric history often overlap with the experience of disability, questions from disability studies resonated with us given how much of our work seeks to unpack the normalizing tendencies of health and medical practice.

Again, COVID-19 had a role to play here: the ableism of the archive itself, with attendant questions of access and elitism, drove us to think through and with a disability studies perspective (see Jirik's essay for a powerful analysis).[16] How could we write ethically about people who had been rendered disabled by both society and history in the middle of a mass-disabling event? It has indeed been enlightening to see historians of health and medicine trumpet the lessons of pandemic histories while simultaneously returning to in-person conferences or seminars from which our disabled colleagues remain excluded. Nicole Lee Schroeder explores some of these tensions in her essay in this volume, and that should give us pause. As historians, we are not immune to bad practices: to hypocrisy or to exploitation of our subject matter for academic gain. In fact, the system we work in encourages and rewards that exploitation. If we are writing about some of the ethical problems with health and medicine in the past, as many of us are, then we need to be authentic about our concerns and self-aware in our own practices.

Now more than ever, history is under attack, and this is for ideological reasons. If history were actually irrelevant, politicians and university administrators would let it die a natural death. But they know there is power in what we write and research and teach and exhibit. Lamenting social justice concerns as presentism in historical work is an abdication of our responsibility as historians, both to the public and to our profession. For this reason, our work must be held to the highest ethical standards we can manage, and we must be self-aware and accountable. We might not always get it right, but we must always try.

***

In this book, we collectively build on the work of other scholars who have explored ethics for historians, especially Susan Lawrence, but also the work of scholars in other fields who have been questioning what it means to be ethical as a researcher, teacher, archivist, librarian, and scholar. This book is *not* intended to be a complete guidebook on "how to be an ethical health historian." In fact, we have deliberately shied away from suggesting any kind of prescriptive formula or "code of ethics," because this book is intended as a first pass, not a last word, on the subject (see the coda for more reflection of this kind). We, the authors and editors, hope to provoke future conversations among health historians about our ethical responsibilities to our subjects, our students, and our publics—and indeed, as a collective, we often disagreed as to what might be the most ethical path through a given dilemma. As editors, we encouraged our authors to ask questions—even if they did not or could not provide fully satisfactory answers to them—in the hope that the readers of this volume will also begin to ask themselves these and similar questions as they approach history.

These essays are thus intended to provide a starting point for discussion and study for historians at every level, especially students and junior scholars learning about the craft of historical research. If students and junior scholars consider their ethical responsibilities as they develop their first research projects, their first syllabi, their first exhibits, and so forth, then, going forward, ethical considerations will be baked into all of their work, a necessary first port of call in the development of health history. Ethical questions should be central to our practice: they should be the foundation for developing historical inquiries and projects, not an optional step, easy to ignore or take lightly. Each of the contributors has posed some of the questions that have animated their own work. Appendix A, at the end of this book, features a number of these questions, and we hope readers find them useful in sparking questions for their own projects.

Placing ethics at the heart of health history may be disorienting. Such a reorientation may preclude some approaches to historical practices that are much beloved or at least taken for granted as normative. The questions raised and approaches proposed in these chapters may make readers uncomfortable with the choices they have made in the past regarding their own research, writing, teaching, or curatorial work. But this kind of productive discomfort, as Cornelia Lambert outlines in her chapter, should not be avoided or shied away from: it is only in confronting the ways in which we have failed in the past—in large ways, in small ways, as individuals, or as a collective field—that we can work together to do less harm.

What, then, does it mean to do *less* harm? The title of this book encapsulates our collective mantra, which emerged from the working group meetings of our first year. The phrase is an acknowledgment that harm has likely been done, if unintentionally—to our historical actors, our students, our publics, our descendant communities, and so on— by health historians, despite our belief in our best efforts and practices. The phrase is also an acknowledgment that doing *no* harm is an impossible goal, albeit one to work toward. An assertion that one can or will "do *no* harm," as many of us who have discussed the modern Hippocratic oath with our pre-health students might suggest, is also an implicit denial that one *could be* doing harm, that one's practices require circumspection and critique. Our oath is therefore to do *less* harm than we have done in the past—by reconsidering our norms and the choices we have made as individual historians and by pushing the discipline to do the same. Our essays demonstrate how, in small and incremental ways, we are attempting to do less harm.

There is no wrong way to read this book. We invite you to choose your own adventure. Perhaps there is a particular question—how to use diagnostic labels in your writing or how to include ethics in graduate instruction, for example—that has been preoccupying you, and a single essay might be a useful starting point to your own approach. Perhaps instead a category of historical practice perplexes you, such as ethical approaches to teaching health history. We invite you to page back and forth, to read individual essays or sections as they might be useful to your thinking. The essays are intentionally short and accessible, to make for both easy reading for historians and approachable teaching material for students.

The book is loosely organized according to a design the working group developed together. Twenty-eight essays on different ethical questions or concerns are united by key topics: "the historian" (which is admittedly ambiguous), "archives and museums," "research," "writing," and "teaching." Because we have been talking and writing together for more than two years, the reader will find many themes or ideas that cross these boundaries and essays that talk to one another—though there are many differences of opinion as well. These essays are thus mimetic of the discussions that produced this volume, loosely scaffolded for the ease of the reader (or use in the classroom).

Part I, "The Historian," focuses on the various ethical issues that affect and shape the way historians themselves approach their work. It raises questions about who has the authority to do historical research in specific communities and how we approach sensitive and personal topics ethically and responsibly. We ask historians to think about their own position in their research, how their work can contribute to reparations for past wrongs, and how historical research can and should be both therapeutic and a source of activism.

Part II, "Archives and Museums," examines the sites that collect, organize, and display the stuff of history. This section, which features essays by museum and archive professionals, as well as other historians of medicine and disability, considers the ways health history is done within these spaces by historians and our ethical responsibilities to these records, objects, images, and the "stuff" of the past. The writers consider, in particular, the ethical responsibilities of historians, archivists, and museum professionals with regard to the care, preservation, and display of human remains and medical records.

Little formal attention has been paid to the ethical responsibilities of the historian undertaking original research, yet in the rapidly shifting legal terrain caused by the passage of HIPAA, there is little guidance on what is "the right thing to do" as opposed to the "legal" things we can get away with. In part III, "Research," we look at the ethical issues raised by genuine, and sometimes disingenuous, attempts to protect records of people, and agencies, in the past. We encourage readers to think about the ethics of the archive itself, to consider questions of violence, power, and authority, and our authority to use and tell the stories of patients in the past. This section also takes a more expansive look at what constitutes an archive and urges readers to "decolonize" their thinking about the methods and approaches we use.

Writing is a central but often underexamined part of the historian's labor. In part IV, "Writing," the authors interrogate aspects of writing the history of medicine and the body encouraging the reader to confront the choices we make as writers from an ethical standpoint. These essays reconsider how we write about people and patients in the past, what aspects of their lives and experiences we discuss, what sources we replicate, who (and why) we cite, and how we share our findings—and our feedback—with historians and the broader public.

Teaching requires choices: we choose which topics to cover, which sources to share with our students, which books to assign them, and what kinds of assignments to require. But our choices are often shaped by the demands of a curriculum, the timing of a semester, and other practical considerations. In part V, "Teaching," the authors challenge the reader to consider the choices made in designing and implementing courses from an ethical perspective. How can we best teach the history of medicine, designing our courses and presenting our material from an ethics-centered approach? These essays offer lessons from the classroom on using ethics to shape health historical pedagogy.

***

Much as doing less harm is an ongoing goal for our authors, and this volume is intended as a starting point for conversation, we also hope that this will only be the first volume of many. These twenty-eight essays, while covering a lot of ground, nevertheless leave some topics by the wayside. There are some topics that we intended to cover but were unable to do so because an author left the project or, in the course of writing, some authors found themselves addressing a different question from their original topic. There are some subjects we wanted to address but could not because the list of essays was already quite long. In other cases, we know that other points of view or ways of thinking about these ethical questions exist and that they have been discussed by scholars in other subfields.

Do Less Harm is a beginning but not an ending. We hope that future follow-up works consider other perspectives, other traditions, and other provocations. We welcome criticism and alternative points of view, and we encourage scholars to contact us if they are interested in joining our community to think about these issues together. Tell us how *you* approach being an ethical historian, and keep your eyes peeled for the next volume, as we work together to do less harm in health history.

NOTES

1. Melvin McLeod, "'There's No Place to Go but Up'—bell hooks and Maya Angelou in Conversation," *Lion's Roar*, January 1, 1998, https://www.lionsroar.com/theres-no-place-to-go-but-up/.

2. Daniel S. Goldberg, "The Shadows of Sunlight: Why Disclosure Should Not Be a Priority in Addressing Conflicts of Interest," *Public Health Ethics* 12, no. 2 (2019): 202–212.

3. See, for example, Anabel Moriña, "When People Matter: The Ethics of Qualitative Research in the Health and Social Sciences," *Health and Social Care in the Community* 29, no. 5 (2021): 1559–1565; and Rhonda M. Shaw et al., "Ethics and Positionality in Qualitative Research with Vulnerable and Marginal Groups," *Qualitative Research* 20, no. 3 (2020): 277–293.

4. Susan C. Lawrence, *Privacy and the Past: Research, Law, Archives, Ethics* (New Brunswick, NJ: Rutgers University Press, 2016).

5. Kacie Lucchini Butcher, "More Questions Than Answers: Interrogating Restricted Access in the Archives," *Journal of the History of the Behavioral Sciences* 58, no. 2 (2022): 223–231.

6. Susan M. Reverby and David Rosner, "Beyond 'the Great Doctors,'" in *Health Care in America: Essays in Social History*, ed. Susan M. Reverby and David Rosner (Philadelphia: Temple University Press, 1979), 3–16; Susan M. Reverby and David Rosner, "'Beyond the Great Doctors' Revisited: A Generation of the 'New' Social History of Medicine," in *Locating Medical History: The Stories and Their Meanings*, ed. Frank Huisman and John Harley Warner (Baltimore: Johns Hopkins University Press, 2004), 167–193.

7. Alexandra Bacopoulos-Viau and Aude Fauvel, "The Patient's Turn: Roy Porter and Psychiatry's Tales, Thirty Years On," *Medical History* 60, no. 1 (2016): 1–18; Roy Porter, "The Patient's View: Doing Medical History from Below," *Theory and Society* 14, no. 2 (1985): 175–198.

8. Ludmilla Jordanova, *History in Practice* (London: Bloomsbury Academic, 2019), 7.

9. Lilia Topouzova, "On Silence and History," *American Historical Review* 126, no. 2 (2021): 685–699.

10. Ann Laura Stoler, *Along the Archival Grain: Epistemic Anxieties and Colonial Common Sense* (Princeton, NJ: Princeton University Press, 2008); Michel-Rolph Trouillot, *Silencing the Past: Power and the Production of History* (Boston: Beacon, 2015).

11. Anton Froeyman, *History, Ethics, and the Recognition of the Other: A Levinasian View on the Writing of History* (New York: Routledge, 2015); Durba Ghosh et al., *Archive Stories: Facts, Fictions, and the Writing of History* (Durham, NC: Duke University Press, 2006); Achille Mbembe, "The Power of the Archive and Its Limits," in *Refiguring the Archive*, ed. Carolyn Hamilton et al. (Dordrecht: Springer Netherlands, 2002), 19–27.

12. James H. Sweet, "Is History History? Identity Politics and Teleologies of the Present," *Perspectives on History*, August 17, 2022, https://www.historians.org/publications-and-directories/perspectives-on-history/september-2022/is-history-history-identity-politics-and-teleologies-of-the-present.

13. Delan Devakumar et al., "Racism, Xenophobia, Discrimination, and the Determination of Health," *Lancet* 400, no. 10368 (2022): 2097–2108; Elise A. Mitchell, "Morbid Crossings: Surviving Smallpox, Maritime Quarantine, and the Gendered Geography of the Early Eighteenth-Century Intra-Caribbean Slave Trade," *William and Mary Quarterly* 79, no. 2 (2022): 177–210; Alexandre I. R. White, *Epidemic Orientalism: Race, Capital, and the Governance of Infectious Disease* (Stanford, CA: Stanford University Press, 2023); Alexandre I. R. White, "Global Risks, Divergent Pandemics: Contrasting Responses to Bubonic Plague and Smallpox in 1901 Cape Town," *Social Science History* 42, no. 1 (2018): 135–158.

14. Hana Burgess, Donna Cormack, and Papaarangi Reid, "Calling Forth Our Pasts, Citing Our Futures: An Envisioning of a Kaupapa Māori Citational Practice," *MAI Journal* 10, no. 1 (2021): 57–67; Linda Tuhiwai Smith, *Decolonizing Methodologies: Research and Indigenous Peoples* (London: Bloomsbury, 2021).

15. Christopher M. Bell, ed., *Blackness and Disability: Critical Examinations and Cultural Interventions* (East Lansing: Michigan State University Press, 2011); Christopher M. Bell, "Is Disability Studies Actually White Disability Studies?," in *The Disability Studies Reader*, 3rd ed., ed. Lennard J. Davis (New York: Routledge, 2010), 402–410.

16. Jarrett M. Drake, "Diversity's Discontents: In Search of an Archive of the Oppressed," *Archives and Manuscripts* 47, no. 2 (2019): 270–279.

PART I

The Historian

CHAPTER ONE

Becoming an Ethical Historian

RICHARD A. MCKAY

Becoming an ethical historian? I can imagine my younger graduate student self reading this title and thinking, *Aha! Reading this will help me to pass the ethics review for my doctorate.* I fear that goal-focused younger me would be slightly disappointed with what now considerably older me has written. For this chapter is not a checklist of rules that, if followed, will guarantee that one's historical work will be "ethical." Rather, this quasi-biographical essay revisits several thorny situations I have encountered from graduate school that deepened my appreciation for ethical historical conduct.[1] These instances of uncertainty arose while applying for ethics review, first as a graduate student and later as a postdoctoral researcher; when approaching prospective participants for interviews and documents; while writing up and publishing research; and when collaborating with scientists, archivists, and documentary filmmakers. Questions I faced included the following: How should I engage with acquaintances and family members of (in)famous historical actors who might be hesitant to share their records and recollections? Could I assert a humanities-centered—as opposed to medical-sciences-dictated—approach to privacy when dealing with highly sensitive topics? And how ought I navigate confidentiality considerations in situations when patient privacy had already been breached? Although I did not fully appreciate it then, these questions, and the answers I found, were also shaped by my positionality as a cisgender gay white man from a middle-class background.

In revisiting these situations, I will suggest that becoming an ethical historian is above all a process—a bumpy, never-ending journey— not a destination. The path may be long and at times difficult to follow.

Nevertheless, navigating a way forward can lead to the deeply satisfying sense of connecting to one's values and of grappling with important questions of truth, trust, and justice. The journey will never be dull. With experience, historians become familiar with general issues arising in research: for example, engaging with historical actors or their descendants, gaining access to records, or untangling privacy and copyright concerns. Novel quandaries will inevitably present themselves, however, entwined with the idiosyncratic particulars of new projects and with evolving debates about the role of scholarship. Drawing from my own experiences and those of other historians, I use the phrase *caring pragmatism* to describe an ethical orientation adopted by historians of medicine and health who care deeply about the historical actors they study and who wish to engage responsibly with the past. By this I mean maintaining a grounding in the principles of ethical research conduct while also embracing each project's unique historical circumstances. These contextual specificities, and often a sense of obligation to the dead as well as the living, can complicate more programmatic ethical approaches often espoused for medical and social science researchers, who typically engage living participants with the aim of creating generalizable knowledge.

Checklists and Hurdles

When I started my graduate research at the University of Oxford, I wanted to understand the rise and spread of the mistaken belief that one man had caused the North American AIDS epidemic. I also wished to recover a "patient's view" for Gaétan Dugas, the so-called Patient Zero, a much-demonized person with AIDS who grappled with the condition long before it acquired its name and widespread recognition. Initially my research felt like a private activity with little risk of creating ripples in the outside world. Gradually, though, my project expanded; I aspired to combine archival research with interviews with scientists, physicians, and members of North America's lesbian and gay communities who had known the Air Canada flight attendant personally (see Sadowsky's chapter on patients' views and competing stakes).

I remember feeling acutely concerned that the university ethics committee held significant power to determine whether my research could proceed, and I approached my application for ethics review with considerable trepidation. I distinctly recall a senior member of my department reinforcing this sense of anxiety, advising me that it was best not to be overly specific in describing how I intended to conduct my research: too many details might unduly raise flags and slow my progress. I also recall a checklist students needed to complete that determined whether their research would qualify for a "light-touch" review. What I didn't know then was that this checklist was heavily weighted toward highlighting potential issues with recruiting patients from the United Kingdom's National Health Service (NHS) or asking interviewees sensitive questions about themselves. Virtually all my research would take place overseas, beyond the jurisdiction of NHS ethics committees, and the sensitive topics I intended to broach pertained to someone *else*, who was deceased, and generally not directly to the interviewees themselves. I can see in retrospect how my honest—albeit circumspect—answers about my project almost miraculously avoided every trigger for a more intensive review. Soon I was free to carry out my research, even though conducting a similar project within the United Kingdom might have required a considerably more thorough review process (see Ramos's chapter regarding institutional review boards).

In terms of training, I recall a single elective session organized for the history faculty's graduate students that introduced different ethical considerations relevant to historians, including scenarios pertaining to scholarly interaction, representing the past, and creating and maintaining the historical record. This session usefully brought ethics to our awareness; however, without also providing a more comprehensive ethics framework onto which to attach the scenarios, it raised far more questions for me than it answered. That year I also received a basic introduction to research ethics during an optional weeklong oral history training course. I left with a keen sense of needing to obtain consent from each interviewee in the shape of a signed copyright transfer form.

These were helpful introductory sessions, but looking back I recognize that my formal ethics training was minimal and entirely elective. Most of my ethical education would be self-conducted (in many ways not much had changed since Susan Reverby's graduate school experience several decades earlier; see Reverby's chapter). I wish I had been pointed then to a best-practice guide like *Ethical Conduct for Research Involving Humans*, produced by the Canadian research councils, that offered a systematic overview of ethical issues in research.[2] I came across this valuable resource much later as a postdoctoral researcher when I was seeking ethics approval for another project, described in more detail later, and as I was processing my Patient Zero interviews for deposit in the British Library. One key insight I would draw from this text echoed something I discovered for myself through conducting and archiving fifty interviews. Rather than thinking of ethical research as a particular hurdle at one moment in time—like passing an ethics review or obtaining a signed consent form—it was far more important to understand it as an ongoing process focused on respectful relationship building (see Pearl's chapter for more on relationships).

Slow-Cooking and Jackhammering

As well as investigating the epidemiologic and social responses to disease outbreaks, I sought to excavate beneath the damaging layers of accumulated media coverage that followed Randy Shilts's popular history *And the Band Played On* (1987), which presented Dugas as a sociopath. Ironically, for a patient who had lived so recently, there were virtually no known documentary sources produced by Dugas himself. Thus, it was important for me to locate people who had known him, who could humanize his story through their personal and professional relationships. I wrote letters to LGBTQ+ publications, spoke to anyone who listened, gave radio interviews, and sent inquiries via an unofficial email list for Air Canada cabin crew. Gradually, responses came back. At the time, I recall thinking that my identifying as gay would help me to establish trust with many in Dugas's extended circle. While this was undoubtedly true, I did not then think more systematically about my positionality and how other facets of my background—my whiteness and middle-class upbringing, for example—almost certainly opened access to other interviews as well.

Increasingly I have adopted the phrase *slow-cooked* to articulate the value of careful historical research—steadily effective, affective labor.[3] In many ways, in-depth historical studies are labors of love. From start to finish, they can easily take many years, risk being underfunded, and may often depend on invisible forms of labor. More positively, at a slower speed I could approach prospective interviewees cautiously, gradually building trust through many moments of contact spread over a long period. Several of Dugas's acquaintances were hesitant to participate, keenly aware of the damaging ways that his story had traveled through the media. Had I faced a more immediate deadline, I suspect that I might have reached overhastily and jeopardized these connections. By contrast, some of the recollections, images, and documents in my book and later in *Killing Patient Zero*, the documentary film based on it, were eventually shared after many years of building trust.

Initially, I did not intend to contact Dugas's family. I knew the media had hounded them in 1987, upon the publication of Shilts's book, and I did not wish to add to their grief. As my research progressed, however, my perspective shifted. I learned that they had been caught unaware when the story broke, since Shilts and his publisher had not contacted them. Although my own research seemed far from their lives in Quebec, my project was growing ever larger. I feared I might unintentionally reenact the traumatic circumstances of decades ago if they learned of my investigations without my first having extended an opportunity for them to contribute. I consulted my supervisors and other experienced historians to gather advice.[4] In late 2007 I wrote a letter explaining my project and inviting Dugas's siblings to contribute. I listed my contact details and explained I would follow up in two months if I did not receive a response. Initially I hoped to relay the message in person through an intermediary in their town of L'Ancienne-Lorette. After a research visit established that this route would feel forced, I posted the letter in early 2008 to a sister for whom I had acquired an address. When I phoned two months later, having not received an answer, I managed to utter an introductory sentence in French, only to have Dugas's sister reply, "Monsieur, I have no desire to speak with you," before hanging up.

From this point, I committed to respecting the family's privacy and steering a wide berth around Dugas's early life. But in 2010, Ray Redford, a former lover and friend of Dugas, shared a letter from him from 1982. This incredible message articulated not only Dugas's illness experience in his own words but also the qualities that many loved ones recalled—his humor, kindness, and care. It seemed vital to quote from this letter. Yet copyright law was resoundingly clear: if I wished to reproduce any quotations, I had little choice but to contact the family again to request their permission.

I felt huge pressure to get this right. I paid a native French-speaking friend to translate an article draft so that the Francophone family members could read it. Not wanting to overreach, I decided not to formally request their permission to quote from the letter in my initial correspondence, saving such paperwork for follow-up discussions, if they arose. I also included a copy of Dugas's 1982 letter, as well as some photos of him that Redford shared with me (see Smith's, Withycombe's, and Reverby's chapters regarding reparatory history and returning to groups and communities being studied).

An anxious month followed. I then received a cautious, uncertain reply by mail, which eventually led to email and telephone conversations. After several exchanges, Dugas's surviving sisters gave me permission to quote from their brother's letter in my forthcoming article and future book. This process in 2013 opened a tentative and respectfully distant line of communication, which later allowed me to alert them that another article, discussed shortly, would likely generate significant media attention, and to serve as a go-between when *Killing Patient Zero* entered production. Given their studious avoidance of media interviews for over thirty years, I advised the documentary filmmakers that Dugas's family would almost certainly not participate. Although I relayed the filmmakers' correspondence, my hunch proved to be correct. The family did not engage, but I was able to keep them abreast of developments throughout the film's production.

I have wondered why Dugas's sisters agreed to my quoting from his letter. I believe the slowness of the historical work gave them ample time to understand my commitment to the project and to presenting a fair-minded depiction of their brother, which in their view served to rehabilitate his posthumous reputation. Also, in the context of many others writing about his life, one sister said, "You were the only one who *asked* for our permission." To me, this remark emphasizes the vital importance of building relationships with respect.

It was only later that I really came to appreciate the importance of "slow-cooking." In 2012, I was invited to coauthor an interdisciplinary study that blended historical and phylogenetic analysis to compare Dugas's HIV-positive blood sample, which he provided in 1983 to the Centers for Disease Control, with other early North American HIV samples. After several rejections and reformulations, our article was accepted and published in *Nature* in 2016.[5] Within forty-eight hours of publication, it featured in nearly 1,000 global news stories, including a front-page *New York Times* article.

During a prepublication telephone press conference, reporters were able to ask questions of Michael Worobey, the study's lead scientist, and me. Most queries concerned the molecular biology, so I had ample time to listen and reflect. The journalists were fascinated by the method the scientists had developed "for recovering viral RNA from degraded archival samples," which they had named RNA "jackhammering."[6] Partly relieved at the respite from attention, and partly irked that my historical work was being taken for granted, I sat in silence in my London home office listening to the international telephone conversation. Suddenly, I was struck by the juxtaposition of "jackhammering"—a phrase suggesting the extraction of valuable content from a resistant source, with little concern for its integrity—with the "slow-cooking" of my own data gathering. In a sense I could excuse the journalists for overlooking my affective labor. I had myself often focused on the extracted content itself, not fully appreciating the emotional holding, relating, and trust building I had undertaken to be granted access to it.

Toward a Caring Pragmatism

While I was writing my Patient Zero book, I also developed a new line of research: the Before HIV project. I would investigate how gay men and other men who had sex with men became a group of interest to public health workers attempting to control sexually transmitted infections in mid-twentieth-century North America and England. Because I identified a voluminous set of relevant patient records held by an NHS organization in London, my project required ethics review and approval from two separate committees—a very different situation from the light-touch review of my doctorate. A university ethics committee for humanities and social sciences would have jurisdiction for research overseas, and an NHS research ethics committee would need to authorize any UK work.

At last I acquired a more comprehensive understanding of the research ethics process from start to finish. I also needed to develop a far more extensive ethics proposal aimed at a new audience: a committee of NHS medical researchers, many of whom would have little familiarity with the questions and approaches favored by historians. Their assumed default study was the clinical trial, so I needed to acquire a new language to describe practices that historians took for granted, translating them into the terminology of qualitative research methods. As I prepared a research protocol, participant information sheets, and consent forms, I became aware of an impending clash of values in terms of ethical approaches to handling sensitive information. Medical researchers preferred to render anonymous any participants who disclosed sensitive information like diagnosis with a sexually transmitted infection. Historians, by contrast, placed far greater emphasis on naming interviewees and recognizing the particularities of their lived experience (see Imada's chapter regarding decisions about naming).

I grappled with this tension for months. During my Patient Zero research, I had found many gay men to be comfortable sharing details of their sexual lives that might typically be deemed "highly sensitive." I eventually concluded that to insist, against someone's wishes, that they remain anonymous for such disclosures could itself constitute unethical conduct. I found it helpful to hold in mind a specific, hypothetical participant: a man in his seventies who had undergone the challenges of coming out in the 1960s and who wished to bear witness to his friends who had died during the AIDS epidemic but, because he had mentioned a single experience of gonorrhea, would be forced to remain anonymous. Eventually I devised a hybrid approach guided by each participant's comfort level. In my project, they could choose to be identified by name, to adopt a pseudonym while leaving unchanged other potentially identifying pieces of contextual information, or to be rendered anonymous, with significant effort on my part to disguise their true identities. This approach was successful, though the additional care required to personalize the degree of anonymity placed much additional work and responsibility on my shoulders as researcher.

Meanwhile, a late-arising concern relating to my "Patient Zero" book draft shook me tremendously. I became worried that substantial parts of my project, with data gleaned from records that documented physicians' and public health researchers' interactions with Dugas, might be considered a breach of patient confidentiality. In other words, my book might exacerbate the same damage that had engendered the notion of Patient Zero in the first place.

To work through this impasse, I consulted a philosopher colleague for a second opinion. He helped me to recognize the flaws in applying a slippery-slope argument to Dugas's case. I also reached out to an experienced historian colleague who suggested I read Susan Lawrence's book *Privacy and the Past: Research, Law, Archives, Ethics*. These reflections and readings on privacy and historical research formed the basis of a substantial new section on ethical questions in my book's introduction. I added a reflection on my own positionality too, partly in response to an anonymous referee's suggestion that this would strengthen my trustworthiness as an author (see Lerner's chapter on positionality).[7] I also decided to adopt a stricter approach to redacting images drawn from documents produced by Centers for Disease Control researchers, to eliminate any possibility that information contained in them might be matched to data already in the public domain and risk the identification of the represented individuals.[8] This decision was not solely based on the particulars of the Patient Zero project. It was also prompted by a pragmatic desire to avoid giving the NHS committee reviewing the Before HIV project any reason to doubt my commitment to ethical conduct. I certainly cared for the safety and privacy of patients, alive and dead, whose experiences I was writing about. But I also cared about my ability to maintain my professional reputation and continue future work. Critics of pragmatism will claim that it can serve as a defense for self-interested behavior. It seems to me, then, that both parts of the phrase, *caring* and *pragmatism*, are required to hold each other in check.

Future Steps

To "agonize"—meaning "to struggle mentally or spiritually, . . . to worry intensely, [or] to struggle to reach a decision"—perfectly describes my state of being, often for long periods, while sitting with ethical quandaries in the absence of formal ethics training (see Imada's and Withycombe's chapters regarding similar unease).[9] I often wrestled with decisions, circling around and around in my head. At times I would feel alone and intensely ashamed for not knowing how to proceed, and then doubly so for not making timely progress. I now see that the following steps usually helped me to break free from this spiral. First, I recognized that I needed to ask for help, and this was OK. Second, I spoke with more experienced colleagues who could make suggestions, reframe concerns, or point me toward resources. Third, I read around to see how other researchers had approached similar questions. And fourth, I learned that historians—juggling, as they do, their ethical responsibilities to a broad range of constituents, living and dead—often carve out unique, caringly pragmatic ways forward (Mahoney describes a similar process for archivists in her chapter).

I can also see, in retrospect, that a relatively narrow set of ethical concerns, whether patient confidentiality or participant anonymization, carried a charge for each project. The discussion group that stemmed from a conference discussion and from which this book emerged has been a wonderful means of broadening my ethical self-education. I am interested to see whether the Do Less Harm collective will evolve to resemble something like the supervision circles seen in professions like psychotherapy and coaching. If we take seriously the notion that becoming an ethical historian is a journey, not a destination, then it follows that this journey will continue throughout our careers. Although one may travel ever further down the road toward the goal of "doing no harm," one never arrives. Time and history march on, and as social formations change, new questions arise and different gaps in awareness become apparent.[10] Those who imagine they know all the answers have ended their journey prematurely.

Increasingly I believe that venturing along this path ranks among the most meaningful aspects of my professional life—even if it means never arriving at a clear endpoint. Ethical conduct is a process, not a declaration or a form. As Shannon Withycombe states in her chapter, the process demands we "keep asking the question, What is ethical history?" Paradoxically, it seems that the best results may come from embracing the likelihood of harm, rather than hoping to eliminate it altogether. In so doing, one could—and should—continuously endeavor to do *less* harm, or better still, strive to *do the least harm possible* under the circumstances. For as any historian will tell you, context matters.

NOTES

1. For their assistance on this journey, I am especially grateful to the participants of the Do Less Harm collective, the people I have interviewed, John Forrester, Simon Szreter, Nick Hopwood, Stephen John, Carolyn Read, John Harley Warner, Naomi Rogers, Gayle Davis, Anne Marie Rafferty, Michael Brady, Katherine Foxhall, Florence Grant, Sue Raymond, Jane McKay, and Theo Raymond. My research projects were supported by a Wellcome Trust Doctoral Studentship (grant number 080651) at the University of Oxford's Wellcome Unit for the History of Medicine, an Economic and Social Research Council Postdoctoral

Fellowship (PTA-026-27-2838) at King's College London's Department of History and Centre for the Humanities and Health, and a Wellcome Trust Medical History and Humanities Fellowship (098705) at the University of Cambridge's Department of History and Philosophy of Science. Oxford's Social Sciences and Humanities Inter-divisional Research Ethics Committee, Cambridge's Ethics Committee for the School of the Humanities and Social Sciences, and the City & East Research Ethics Committee, part of the UK's National Research Ethics Service, reviewed and approved the research discussed in this chapter.

2. Canadian Institutes of Health Research, Natural Sciences and Engineering Research Council of Canada, and Social Sciences and Humanities Research Council of Canada, *Tri-council Policy Statement: Ethical Conduct for Research Involving Humans* (Government of Canada, December 2022), https://ethics.gc.ca/eng/policy-politique_tcps2-eptc2_2022.html.

3. This section draws on two presentations: Richard A. McKay, "Patient Zero, Affective Labour, and the Value of Historical Research" (Elias E. Manuelidis Memorial Lecture in the History of Medicine, Yale University, February 18, 2019); and Richard A. McKay, keynote lecture introducing *Killing Patient Zero* (Fadoo Productions, 2019) at the KDocs Film Festival, Vancouver, Canada, February 22, 2020, YouTube video, 12:50, posted August 29, 2020, https://youtu.be/YkX2qPDCMQo.

4. Linda Kerber was particularly helpful in suggesting sensitive approaches.

5. Michael Worobey et al., "1970s and 'Patient 0' HIV-1 Genomes Illuminate Early HIV/AIDS History in North America," *Nature* 539 (November 3, 2016): 98–101.

6. Worobey et al., 98.

7. Richard A. McKay, *Patient Zero and the Making of the AIDS Epidemic* (Chicago: University of Chicago Press, 2017), 32–39.

8. See figures 2.7 and 2.8 in McKay, 127–128.

9. *Oxford English Dictionary*, s.v. "agonize (v.), sense 4," September 2023, https://doi.org/10.1093/OED/6193424609.

10. I address one of my own gaps in awareness, in relation to race, racism, and whiteness in my Patient Zero work, in Richard A. McKay, presentation in "On the Margins: Epidemics and the Disenfranchised," Pandemic Histories, online seminar, Canadian Society for the History of Medicine, June 15, 2020, YouTube video, 1:30:46, posted July 18, 2020, https://youtu.be/3paQ7cpCLS8.

CHAPTER TWO

Positionality and the History of Medicine

BARRON H. LERNER

In a sense, I am an unusual choice to write a chapter on positionality in the history of medicine. *Positionality*, a term first popularized in anthropology, can be defined as an exploration of a researcher's perspective with respect to the topic being discussed. That is, part and parcel of the work itself is an acknowledgment of who the researcher is and their perspectives.[1]

When I was a graduate student in the history of medicine in the 1980s and 1990s, no one mentioned positionality. In retrospect, however, it is clear that the ideas it conveyed were in the air. For decades, the history of medicine was written by physicians about physicians, focusing predominantly on the achievements of the white men who dominated the medical profession. But beginning in the 1970s, social historians—more likely to have PhD than MD degrees—began researching and writing medical history. These scholars argued that previously neglected individuals, including nurses, midwives, patients, and research subjects, needed to be part of medical history, as did discussions of race, class, and gender. The new social historians argued not only that the old history had been exclusionary but also that it was "Whig history," which falsely saw the history of medicine as a relentless march of progress. The positionality of the social historian had challenged that of the physician-historian.

While this new scholarship questioned and expanded on aspects of traditional medical history, it did not fundamentally upend the basic tenets of the field. But now, reflecting an increased interest in positionality, a more substantial historiographic change is afoot. Current writing in the field, largely but not exclusively focusing on the history of race and racism, has argued that exploitation of certain groups in society has been a fundamental, structural part of medicine and health care for centuries. As such, focusing on the roles played by concepts such as racism, violence, and colonialism is not only an essential way to better understand all medical history but perhaps a scholar's ethical duty.

Related to this radical new view of medical history is a revisitation of the concept of presentism. In contrast to positionality, we discussed presentism all the time during my graduate school days. There are many forms of presentism, including Whig history, mentioned earlier, that reflexively view the past as a prelude to a more enlightened present. But for the purposes of this chapter, I am using the term *presentism* to refer to scholarship that assesses past events using any type of a modern ideological lens. Doing so, my professors argued, was unequivocally bad history. Past events needed to be understood exclusively in the context of their particular time periods, thus enabling the historian to learn why specific historical figures had made specific choices. But the new focus on positionality has led to a reconsideration of presentism. Perhaps, as Courtney Thompson and Kylie Smith argue in the introduction to this book, presentism is an inevitable—and desirable—component of any historical research. Instead of rejecting it, they believe, we should embrace it and explore how best to harness it to produce better historical scholarship.

As someone who had given little thought to positionality but who was trained to avoid presentism at all costs, I found writing this chapter to be a challenge that both caused me discomfort and led me to rethink some of my most cherished beliefs about history. Focusing primarily but not exclusively on the United States, this essay will explore the recent epistemological challenges to more traditional medical history.

Great Man History

An excellent example of great man medical history is Harvey Cushing's 1925 *The Life of Sir William Osler*.[2] Osler had been the most famous physician in North America, a founder of the Johns Hopkins School of Medicine and author of humanistic writings about medicine.[3] Simply by following Osler's life, Cushing put his readers at the cutting edge of medical discoveries and introduced them to other esteemed physicians. And it made perfect sense for Cushing to have authored the book. Himself a famous neurosurgeon, he had worked closely with Osler.

When the book won the 1926 Pulitzer Prize for Biography or Autobiography, it only validated the importance of such a narrative for understanding medical history.[4] This is not to say that there were no precursors of social history. For example, Henry Sigerist was a Swiss physician and historian who explored the social and political aspects of medical history.[5]

The changes that occurred in the history of medicine starting in the 1970s drew on broader developments in American society. Social change, such as the civil rights movement, anti–Vietnam War protests, and second-wave feminism, rocked the country. Meanwhile, scandals within medicine, such as the Willowbrook hepatitis experiments and the Tuskegee Study of Untreated Syphilis in the Negro Male, had called into question the ethics of the great white doctors whom medical history had so long celebrated.[6]

Out of this milieu emerged an edited volume, *Health Care in America: Essays in Social History*, in 1979. In an introductory essay, "Beyond 'the Great Doctors,'" coeditors Susan Reverby and David Rosner argued that the predominant focus on white male doctors ignored a vast number of experiences and events.[7] Chapters focused, instead, on topics such as the Ladies Physiological Institute of the 1850s and the social mores of obstetrics in the mid-nineteenth century. The entrance of social historians introduced not only new historical actors but dramatically different ways of understanding historical topics such as diseases, medical technologies, and public health campaigns.

Perhaps inevitably, the arrival of this new perspective caused tension. Certain physician-historians were very critical, including one who termed the new social history "medical history without medicine." Social historians returned the favor, suggesting the traditional physician-dominated narratives were "medical history without history."[8] But certain scholars, including clinicians who obtained advanced degrees in history, helped to bridge the divide, writing social history about topics likely to interest practitioners. I was one such individual.

Just as great man history had reflected the interests of its authors, so too did social history. Looking back, it is clear that this was positionality too. Many early social historians of medicine were feminists who wrote articles and books on the history of women's health.[9] Other historians, reflecting contemporary concerns about human experimentation and institutionalization of the mentally ill, explored how social and cultural factors had influenced medical diagnoses and treatments.[10] Historians of public health, in particular, were often unabashedly activist, unearthing stories of corporate America promoting unhealthy—and even deadly—products in search of profits.[11]

But there were gaps in the social history of medicine as well. In particular, perhaps reflecting the relative absence of historians of color in the field, few works explored race and racism in medical history. For good or bad, the high-quality works that did exist—by historians such as Darlene Clark Hine, Vanessa Northington Gamble, Keith Wailoo, Evelynn Hammonds, and Rebecca Herzig—regularly appeared in survey-type courses to ensure that race was "covered."[12] Unfortunately, this scholarship may have led some historians of medicine to downplay or even ignore the important role that race also played in the stories they were telling. It could be argued that I was one such historian, having authored a book on the history of medical ethics featuring two white, male physicians—my father and me—and dealing only tangentially with race. Such gaps have energized calls for the reparatory history advocated by Kylie Smith and others in this book.

For a graduate student in history in the 1990s, presentism was a cardinal sin. Professors teaching the history of medicine, a topic with potentially enormous modern relevance, beseeched their students not to interpret historical findings through a modern lens. The lesson of the Tuskegee study, for example, was not that its organizers were far more unethical and racist than modern investigators. Rather, they were imperfect people who made bad choices, continuing to research poor Black men with syphilis as opposed to being concerned about their worsening illnesses and deaths.[13] Historians needed to study what factors had caused them to do this. Most historians of medicine would have admitted that modern issues influenced their research and that it was acceptable to draw relevant conclusions from one's work, but in an era before positionality, they never would have announced this. Rather, there was a tacit understanding that scholars would use their archival findings to produce as unbiased a history as possible.

Epistemological Challenges

But what if traditional approaches to the history of medicine were incomplete or—even more worrisome—inaccurate or misleading? That is, what if the usual concerns of medical history actually obscured the ways in which events had occurred or knowledge was obtained? These types of epistemological questions began to be asked in the 1990s and have subsequently become more common, especially due to the work of scholars looking at slavery and racial injustice.

Building on the notion of creating different lenses to revisit historical knowledge, historian Deirdre Cooper Owens's book *Medical Bondage: Race, Gender and the Origins of American Gynecology* focuses in part on the "mothers of gynecology," her term for enslaved women who underwent gynecological interventions without any ability to decline.[14] What was "learned" from these "medical superbodies" were false, racialized beliefs about Black women, such as their supposed hypersexuality and immunity to pain, that (unfortunately) persist today. Exploring Jamaica and South Carolina from 1780 to 1840, historian Rana Hogarth similarly found that physicians constructed racially based biological differences and medicalized Blackness, choices that had long-lasting impact on the evolution of medical beliefs.[15]

Cooper Owens also notes a surprising finding: slave women, including some participants in the infamous vesicovaginal fistulae experiments of gynecologist J. Marion Sims, also worked as nurses and midwives. In this manner, they actively participated in the production of medical knowledge.[16] As such, these previously invisible women were now revealed to be important figures in the history of medicine. And among the forgotten historical figures were not only enslaved people but also others involved in health-related interactions. For example, as Ryan Johnson and Amna Khalid have written, subordinate and intermediary agents, including health officials, nurses, and midwives, were central to the practice of public health in colonial outposts of the British empire.[17] Another previously ignored group that helped advance epidemiological knowledge was injured soldiers. According to Jim Downs, historians needed to "flip the script," making these types of individuals—as opposed to the physicians and researchers in charge of their medical care—the protagonists of their stories.[18]

The blatant oppression that Cooper Owens's and other historical subjects had experienced raised another epistemological point: Hadn't medical historians downplayed or disregarded the roles played by subjugation, control, and even violence when recounting the history of Western medicine? In other words, weren't these issues not just occasional topics in the history of medicine but rather intrinsic to understanding the basic evolution of clinical practice and patient experiences?

New scholarship unearthed similar themes in the history of medical education. For example, in *Masters of Health: Racial Science and Slavery in U.S. Medical Schools,* historian Christopher Willoughby shows that both Southern and Northern medical schools in antebellum America taught white supremacist medical science—theories that emanated not from science but from racist medical stereotypes.[19] As with Cooper Owens's gynecologists, medical students embraced these ideas—such as the notion that Black people were different anatomically from whites—and later taught them to their own students.

Revisiting Presentism

In sum, the new scholarship argued that historians of medicine too often ignored how disadvantaged populations were fundamentally involved in—and in some cases responsible for—how medicine evolved over time. This concept, in turn, raised fascinating questions about the issue of presentism. The sacrosanct view that modern understandings should not influence historical findings had come under fire.

An excellent example of this argument is the idea of structural violence, which argues that first colonialism, and then imperialism, put certain individuals and populations "in harm's way." Although this term has existed since the 1960s, it received renewed attention after police murdered George Floyd in 2020. And while much attention has been paid to racism in law enforcement, a similar argument has been made in the case of the history of medicine: "Racist policy and practice have . . . been integral to the formation of the medical academy in the USA."[20]

Traditional historians of medicine might feel uncomfortable using this idea as a starting point for historical scholarship. After all, if you begin with such a passionate belief, won't it potentially influence your findings, encouraging you to pick historical anecdotes and concepts that support your worldview? Put another way, doesn't this sort of work serve activism but not necessarily history?

Related to this argument is the debate over seeing historical figures as "of their time." To reiterate, critics of presentism have inveighed against identifying antecedents of modern concepts—such as structural violence—in historical events. This would be ahistorical, suggesting that unappealing historical figures should have known that what they were doing was wrong. The proper strategy is to look at their choices in their own historical context and then bring these findings and "lessons" to the present.

But more activist historians now view this notion of presentism as "intellectually shortsighted" and believe that it potentially excuses past racism and violence. Dawn Peterson and Laurel Clark Shire make this point regarding historical understandings of American president Andrew Jackson. Rather than explaining away Jackson's genocidal actions against Indigenous people as emblematic of the racist and unenlightened times in which he lived, his actions should alternatively be seen as "part of a national racist project that sought profits from the theft of land and labor by establishing and maintaining white supremacy."[21] Given this history, it has been argued that scholars have an ethical duty to produce antiracist and anticolonial scholarship that centralizes oppressed people's struggles for freedom and autonomy.[22]

A similar case can be made for medical history. In a 2020 article, historians Ayah Nuriddin, Graham Mooney, and Alexandre White review a series of historical events, including compulsory eugenic sterilization before World War II, the Tuskegee study, and the harvesting of cancer cells from Henrietta Lacks, as past examples of structural violence. That is, this modern term can provide a unifying historical umbrella for understanding past events and linking them to present health concerns.[23] An especially powerful argument for making "personal connections with the historical actors we study" is Cooper Owens's jarring account of her own experience as a Black medical superbody, being subjected to a painful fertility procedure without anesthesia, after having written about Sims's operations.[24]

Having said this, by no means are modern historians of medicine merely looking back in the historical record to find support for modern causes. Cooper Owens's unexpected finding about slave women participating in gynecological experiments complicates her narrative, as does Nuriddin's work on eugenics, which describes how certain Black Americans actually embraced eugenics as a way to fight social injustice. Yes, it may be incorrect to excuse past behaviors as "of their time," but historians are still obligated to carefully report and historicize their findings.[25]

These issues came to a head in August 2022 when American Historical Association president James Sweet wrote a column suggesting that activist historians were plumbing the historical record for examples of events that supported their causes—that is, being presentist.[26] The response to Sweet was fast and furious, with scholars arguing, among other things, that history is always political. That is, using terms like *structural violence* and *systemic racism* as a way to explore the past is no different from the practices of generations of earlier historians who "whitewashed national myths" and systematically ignored the contributions of disadvantaged populations to historical events and knowledge.[27] Sweet later issued an apology.

Conclusion

I did not "appear" in the first draft of this chapter. The history of medicine that I have written over thirty-plus years almost always addresses only historical figures—not me. As one historian once said, "History exists to learn about *other* people." Although I pondered the possible disconnect of a chapter on positionality in which the author does not announce his or her positionality, I felt that anyone interested in who I am—a white, male physician-historian who is a past president of the American Association for the History of Medicine—could easily figure this out on Google or my website. But my coeditors and coauthors disagreed, arguing that an essay that simultaneously announced my privilege and grappled with positionality would send an important message about the value of new scholarship in medical history.

Following in the footsteps of the social historians who encouraged history "from the bottom up," modern historians of medicine are offering another historiographic challenge. The history of medicine, they argue, has consistently been characterized by violent and racist oppression of patients, research subjects, and other vulnerable groups. Not only should this fact be acknowledged, but it should provide a basis for future historical research. And while this chapter has focused mostly on race and racism, other scholars, such as historians of disability (see Jirik's chapter in this book), are introducing other lenses for reinterpreting traditional histories of medicine.[28]

So, having authored a chapter on positionality, how do I feel about it? As most historians would agree, the thoughtful evolution of a field is to be embraced. Merely doing things the way one was taught is often shortsighted. And the innovative scholarship cited in this chapter has genuinely gotten historians of medicine to think in new ways about the past—and how it should be explored.

But it would be a mistake to end a chapter on positionality without reminding readers that it may be harder to historicize the present as opposed to the past. The decades-old admonition to avoid presentism that Sweet sought to champion was not necessarily "right" or "wrong" but rather a historical choice made by scholars of a particular era. Similarly, more recent efforts to champion positionality and to think of presentism as helpful to the historical project are themselves choices—ones that future historians can and should study.

NOTES

1. Andrew G. D. Holmes, "Researcher Positionality—a Consideration of Its Influence and Place in Qualitative Research—a New Researcher Guide," *Shanlax International Journal of Education* 8, no. 4 (2020): 1–10. I would like to thank all of my coauthors in this volume for their helpful feedback, and especially Jonathan Sadowsky, who was a particularly valuable sounding board.

2. Harvey Cushing, *The Life of Sir William Osler* (Oxford: Clarendon Press, 1925).

3. Howard I. Kushner, "Medical Historians and the History of Medicine," *Lancet* 372 (August 30, 2008): 710–711.

4. Beth Linker, "Resuscitating the 'Great Doctor': The Career of Biography in Medical History," in *The History of Poetics and Scientific Biography*, ed. Thomas Söderqvist (Aldershot, UK: Ashgate, 2007), 221–239; Barron H. Lerner, "Great Doctor History: A Personal Journey," *Bulletin of the History of Medicine* 92, no. 1 (2018): 55–77.

5. Elizabeth Fee and Theodore M. Brown, eds., *Making Medical History: The Life and Times of Henry E. Sigerist* (Baltimore: Johns Hopkins University Press, 1997).

6. James H. Jones, *Bad Blood: The Tuskegee Syphilis Experiment* (New York: Free Press, 1992); Sydney A. Halpern, *Dangerous Medicine: The Story behind Human Experiments with Hepatitis* (New Haven, CT: Yale University Press, 2021).

7. Susan M. Reverby and David Rosner, "Beyond 'the Great Doctors,'" in *Health Care in America: Essays in Social History*, ed. Susan M. Reverby and David Rosner (Philadelphia: Temple University Press, 1979), 3–16.

8. David Rosner, "Tempest in a Test Tube: Medical History and the Historian," *Radical History Review* 26 (1982): 166–171.

9. Judith W. Leavitt, *Brought to Bed: Childbearing in America, 1750–1950* (New York: Oxford University Press, 1986); Leslie J. Reagan, *When*

Abortion Was a Crime: Woman, Medicine and Law in the United States, 1867–1973 (Berkeley: University of California Press, 1998).

10. David J. Rothman, *Strangers at the Bedside: A History of How Law and Bioethics Transformed Medical Decision Making* (New York: Basic Books, 1991); Joel Braslow, *Mental Ills and Bodily Cures: Psychiatric Treatment in the First Half of the Twentieth Century* (Berkeley: University of California Press, 1997).

11. Robert Proctor, *Golden Holocaust: Origins of the Cigarette Catastrophe and the Case for Abolition* (Berkeley: University of California Press, 2012); Gerald Markowitz and David Rosner, *Lead Wars: The Politics of Science and the Fate of America's Children* (Berkeley: University of California Press, 2014).

12. Darlene Clark Hine, *Black Women in White: Racial Conflict and Cooperation in the Nursing Profession, 1890–1950* (Indianapolis: Indiana University Press, 1989); Vanessa Northington Gamble, *Making a Place for Ourselves: The Black Hospital Movement, 1920–1945* (New York: Oxford University Press, 1995); Keith Wailoo, *Dying in the City of the Blues: Sickle Cell Anemia and the Politics of Race and Health* (Chapel Hill: University of North Carolina Press, 2001); Evelynn M. Hammonds and Rebecca M. Herzig, eds., *The Nature of Difference: Sciences of Race in the United States from Jefferson to Genomics* (Cambridge, MA: MIT Press, 2009).

13. Susan M. Reverby, "Ethical Failures and History Lessons: The U.S. Public Health Service Research Studies in Tuskegee and Guatemala," *Public Health Reviews* 34, no. 1 (2012): 1–18.

14. Deirdre Cooper Owens, *Medical Bondage: Race, Gender and the Origins of American Gynecology* (Athens: University of Georgia Press, 2017).

15. Rana A. Hogarth, *Medicalizing Blackness: Making Racial Difference in the Atlantic World, 1780–1840* (Chapel Hill: University of North Carolina Press, 2017).

16. Cooper Owens, *Medical Bondage*, 42–72.

17. Ryan Johnson and Amna Khalid, eds., *Public Health in the British Empire: Intermediaries, Subordinates, and the Practice of Public Health, 1850–1960* (New York: Routledge, 2011).

18. Jim Downs, *Maladies of Empire: How Colonialism, Slavery and War Transformed Medicine* (Cambridge, MA: Harvard University Press, 2021), 7.

19. Christopher D. E. Willoughby, *Masters of Health: Racial Science and Slavery in U.S. Medical Schools* (Chapel Hill: University of North Carolina Press, 2022).

20. Ayah Nuriddin, Graham Mooney, and Alexandre I. R. White, "Reckoning with Histories of Medical Racism and Violence in the USA," *Lancet* 396 (October 3, 2020): 949–951, quote on 949.

21. Dawn Peterson and Laurel Clark Shire, "A 'Man of His Time' and the Methods of White Supremacy: #SHEAR2020 and the Plenary Debacle," *Panorama*, August 7, 2020, https://thepanorama.shear.org /2020/08/07/a-man-of-his-time-and-the-methods-of-white-supremacy -shear2020-and-the-plenary-debacle.

22.   Arinn Amer, "Remembrance of Things Not Yet Past: A Report from 'Difficult Histories/Public Spaces: The Challenge of Monuments in NYC and the Nation,'" Gotham Center for New York City History, September 18, 2018, https://www.gothamcenter.org/blog/remembrance-of-things -not-yet-past-a-report-from-difficult-histories-public-spaces-the-challenge -of-monuments-in-nyc-and-the-nation.

23.   Nuriddin, Mooney, and White, "Reckoning with Histories."

24.   Cooper Owens, *Medical Bondage*, 123–126.

25.   Amer, "Remembrance of Things"; Ayah Nuriddin, "The Black Politics of Eugenics," *Nursing Clio* (blog), June 1, 2017, https://nursingclio .org/2017/06/01/the-black-politics-of-eugenics.

26.   James H. Sweet, "Is History History? Identity Politics and Teleologies of the Present," *Perspectives on History*, August 17, 2022, https://www .historians.org/perspectives-article/is-history-history-identity-politics-and -teleologies-of-the-present-september-2022.

27.   Priya Satia, "The Presentist Trap," *Perspectives on History*, September 7, 2022, https://www.historynewsnetwork.org/article/history-as-love -and-the-presentist-trap-responses-.

28.   Aparna Nair, "Epilepsy and Me," Wellcome Collection, accessed November 13, 2024, https://wellcomecollection.org/series/XjlgkREAACUA _s3s; Jaipreet Virdi, *Hearing Happiness: Deafness Cures in History* (Chicago: University of Chicago Press, 2020); Katrina N. Jirik, "Parents, Superintendents, and Lawmakers in the Creation of Institutions for the Feeble-Minded, 1876–1916," *Pennsylvania History: A Journal of Mid-Atlantic Studies* 89 (2022): 412–428.

CHAPTER THREE

Reparatory History

KYLIE M. SMITH

Historical knowledge has always been necessary in building the case for reparations, especially in relation to slavery, genocide, or theft of native land. But what does it have to do with being an ethical historian of medicine or health care? The very idea of reparation assumes that something bad has happened, some foul deed or collective evil, which some may argue medical and health care practices are not guilty of. But as the title of this volume suggests, we start from the assumption that medical and health care practitioners, systems, and the field's historians have indeed at times done harm.

Over the last decade, a new generation of historians have sought to document the harm done in and by the history of medicine, particularly in the context of US slavery.[1] These scholars have taken a more "social history" approach in order to try to shift the focus away from the "great white doctor" narrative and consider the impact of other actors, as well as include the perspectives of patients and communities themselves. So what makes a reparatory approach different from this new critical history? In this chapter, I argue that reparatory history offers a potentially ethical framework for understanding how to approach our work as historians of health and medicine that is aimed explicitly at creating accountability. This quest for accountability, I argue, can guide the subject of our work, shape our own approach as historians, and inform what we do with our research.

In the next section I define and explain what is meant by the term *reparatory history* and consider why it matters for the history of health care. I then give a case study of how I am using it in my own work on the history of psychiatry in the US South as well as dealing with some of the ethical challenges posed by taking this approach. The main argument I make here is that we need to take seriously the harm that has been done by racist, ableist, and sexist medical practices and that, as historians of health and medicine, in the current moment we need to take an ethical stance that is attuned to social justice. But reparatory history is not without challenges—we must still proceed with caution. Declaring something a reparatory history does not automatically make it ethical. In fact, this approach asks us to be even more careful about how we write histories that are designed to expose past evils, especially in the field of health and medicine, where the result could easily be retraumatization and exploitation.

What Is Reparatory History?

As the name suggests, the idea of reparatory history relates most obviously to the long call for reparations or compensation in relation to slavery in both the United Kingdom and the United States that dates back to abolition and the Civil War, but also in relation to the Holocaust and South African apartheid, as well as Indigenous dispossession and genocide.[2] There are two main principles here: that things have been done for which governments, organizations, and individuals need to be held accountable; and that those are things for which a debt (usually some kind of monetary redress) is owed.

Most of the scholars of reparatory history come from either slavery or Holocaust studies, and they are overtly concerned with compiling and presenting historical evidence that can be used in making the case for financial reparations.[3] But they also make some more philosophical claims that touch on recent debates about the so-called presentism of justice-oriented history. That is, reparatory historians argue against a progressive narrative of history—they do not see a rupture between past wrongdoings and current social and political life. Rather, they see continuities between past and present, and so there is an unapologetically political nature to this kind of historical writing. Reparatory historians take an overt stand that historical research and writing should have a justice orientation and that this is an ethical obligation for history as an enterprise, especially in our current moment.

In his article "A Reparatory History of the Present," David Scott writes, "In the face of the dead ends of racial justice that define our present, it is reparatory history that ought to command our attention."[4] By "reparatory history," Scott means the type of history that "is concerned with those historical evils and injustices that remain unrepaired in the present, whose wrongs continue to disfigure generations of human lives," and that seeks to "reconstruct these evil and unjust pasts in ways that potentially enable us to rethink the moral responsibility that the present owes in respect of them."[5] Scott is largely referring here to the "big evils" of slavery and the Holocaust, and of course, history is central to reparations for these events: historians are the ones who create the material by which it is determined "what story of the past is being linked to what demand in the present and what imagination of the future."[6] Calculating the financial compensation that is undoubtedly owed in terms of lives lost, land stolen, and wealth forestalled is one aspiration of the reparations movement. In relation to apartheid, truth and accountability are central to reconciliation and to the very possibility of a functioning society.[7] In the postcolonial world, reparatory history focuses on questions of how anticolonial movements can be more than oppressive nationalistic ones.[8] Reparatory history is reparatory because it exposes past actions that a state can be forced to address in terms of the present and the future—it is reparatory in that it seeks justice and a different way of being from now on.[9]

Case Study: Reparations in the History of Psychiatry

But what does all of this have to do with ethics in the history of medicine or health care? Here I want to consider how I use these ideas in my own work in the search for an ethical framework, and how they have shaped both my approach and the outcomes of my research and writing. I also consider some of the ethical implications of doing reparatory history, in terms of what we owe both the dead and the living.

In January 2021, after a long summer of racial reckoning, the American Psychiatric Association published an apology for its role in "the support of racism in psychiatry."[10] This vague statement gave no real specifics about what this "support" looked like, nor did it really account for the internal racism of psychiatric thought and practice itself. My work on psychiatric hospitals in southern US states like Georgia, Alabama, and Mississippi is deliberately designed to address this silence and to hold psychiatry accountable.

In this research, for example, I have found a long history of neglect, abuse, and violence toward Black patients that was sustained and reinforced by psychiatrists who promulgated false beliefs about differences between Black and white emotional and psychological processes. These beliefs created and justified systems that segregated Black patients from white ones, alienated them from their families, and forced them to perform hard labor under the guise of therapy.

The damage that was done to families and communities matters not just because of what was done to people in the past but also because of what is still being done to people in the present. If these hospitals were closed in recognition of their horrors, the ideas that sustained them have not been eradicated, merely displaced to other terrains like the carceral system. And the false science used to argue that race is a biological category rather than a complicated social construction with real-world consequences continues to permeate health research today. Reparatory history gives me a framework through which to explore the continuities of these practices and to think about how I can approach my research material in an ethical manner.

Writing this history has been a challenging process: I am a white woman not from the United States, and this position has often given me pause. As Barron Lerner explores in this volume, considering our position as people within the work we are doing should be considered as the first step toward some kind of ethical practice. My position is also an overtly political one. I am quick to acknowledge that I do not seek to document a balanced view of the history of psychiatry. In this volume and elsewhere, Jonathan Sadowsky has laid out some of the tensions for the historian of psychiatry who is also writing for the living.[11]

In my study, I overtly focus only on Black patients, and I seek to situate what happened to them within the context of Southern white supremacy in the mid-twentieth century. There is very little that is redeemable in this history, and I take my lead from reparatory historians like David Scott to argue that a reparatory history of psychiatry is one that attempts to document practices and goes beyond apologizing for "a few bad apples" or making vague gestures at identifying "structural inequities." It is a history that documents the ways that psychiatry itself has actively shaped racist ideas about the Black personality and subjectivity both within and beyond the clinical encounter, ideas that are foundational to the way that American white supremacy continues to manifest and flourish. Psychiatry is not slavery, and the asylum is not the plantation, but there are links between them in the same way that there are links between the plantation and the prison.[12] A reparatory history of psychiatry or health care should not shy away from this fact but should be the link between the past and present, in an analysis of both structural systems and the ideas and practices that underpin them, especially when they are foundational to enduring ideas and practices of race and racism.

Beyond positionality and politics, there are other ethical aspects to consider in the pursuit of reparatory history. The first is what we owe to the dead and how we recount narratives of evil without reproducing people as victims. I take my lead from Saidiya Hartman here, who asks the question, "How and why does one write a history of violence?"[13] Hartman actively encourages reparatory history, arguing that we must write our histories without fear or favor, but she warns that this approach does not give us carte blanche to re-create trauma or exploit our subjects, and that we must try to imagine what life was really like for the people who suffered and resisted.[14] This is a complicated enterprise when it comes to the history of psychiatry because of the nature and politics of the records that have been preserved.[15] There is also an important issue to consider in relation to the outcome of our work, and how we avoid benefiting from someone else's suffering while putting our work to use with the goal of reparation.

There are a number of ways I have engaged with these tensions and tried to enact a reparatory approach. First, I made a decision to share archival findings as much as possible with families and descendants of the patients whose traces I find and ask them to guide me in what they would like to see published. One family's story has become the prologue to the book I am working on, and they have had the final say over the written words. But this process was not without risks or ethical questions. As historians, we tend to think that everything we find in the archives is ours to use, but when I found Mrs. Hurston's letter, my thoughts turned to questions of reparations and what sort of responsibility I owed to her living family. I wanted to reach out to them, but I needed to be ready for the family to tell me they did not want me to use their story, and to leave it out of the book if that was what they wanted. Having the goal of reparations or accountability did not give me the right to trample all over other people's privacy. In this case, the Hurstons were proud of their mother and brother and understood that publication of their story could be used to hold the State of Mississippi to account. But it also comes at a risk to them and requires a willingness to make their story public.

The public-facing nature of reparatory history requires careful ethical consideration. In an attempt to mitigate some of the harm that can come from my publishing and earning royalties from this work, I have deliberately pursued open-access publishing. The book that is emerging from all of my research will be published by the University of North Carolina Press as both a paper volume and an open-access "digital enhanced monograph" on the open-source publishing platform Manifold.[16] The goal of projects like these is to make the work we do as historians publicly available and accountable to the people we serve. For me, it is also accountability to the people I am writing about, who have shared their lives with me or who have been silenced in the archives. My goal with a more public-facing approach is to ameliorate some of the potential for exploitation and to create a platform for future community memory and engagement. Of course, that process is also supported by publication subvention funds from my privileged university and philanthropies like the Mellon Foundation. Open-access publishing should not require financial underwriting by elite institutions, as this simply reiterates and re-creates privilege and continues to silence the stories that need telling the most. While I will be donating any royalties I do earn to community mental health services in Georgia, Alabama, and Mississippi, real reparations would be direct accountability, financial or otherwise, from the state systems and from the American Psychiatric Association, which condoned and supported racist practices in the first place.

Racism in the History of Health Care and the Impossibility of Repair

At the same time, as all the scholars I have cited point out, there is an impossible paradox at the heart of a call for reparations, especially if one is thinking about futurity, and that is the reality that reparations themselves, financial or memorial, are not the end of the story. They do not undo the original evil; they cannot bring back the lives lost or the land stolen. Material forms of reparation cannot be used as a form of closure; there can be no closure because that would mean forgetting. Closure, in that sense, would be foreclosure, used to erase the past so that it can no longer inform the present or future. But some things, some evils, are beyond repair.[17]

Is the history of medicine and health care really one of those evils, though? The common thread throughout all the reparatory history sources I have cited in this essay is the concern with racism. Slavery, apartheid, the Holocaust, these are all manifestations of the great evil of racism and its siblings, colonialism and capitalism.

It is this underlying thread that links past evils to the present, and this is also the case in the history of health, medicine, and the human sciences. For too many years, our field has tended to leave unchallenged the racist ideology that Western medical knowledge is itself built on. New work that is informed by critical race theory, disability justice, and Black feminist thought has begun to challenge this hegemony through a detailed and painstaking analysis of, for example, the ways that plantation management was deeply foundational to the professionalization of medicine, or the ways that nursing care was implicated in the colonial project.[18] These are reparatory histories because they seek to make clear the threads that bind the past to the present and make explicit calls to new futures and the imagination of an antiracist medicine.[19]

If the Black Lives Matter movement and the COVID-19 pandemic revealed to public health and policymaking officials the glaring structural inequities in health care systems, some historians were already only too aware of the underlying racist assumptions that reify disparities as an effect of personal responsibility and a nonexistent racial biology. When one professional organization after another professed "regret" for past mistakes or discriminations, reparatory historians issued a collective groan of frustration at the lack of deep engagement with how current practices themselves replicate past mistakes and continue to sustain themselves on the lives and bodies of people of color.

In the words of Christina Sharpe, modern health care systems continue to operate "in the wake" of the plantation, and revisionist histories and half-hearted statements do nothing to change these deep-seated foundational ideologies.[20] As historians of medicine and health care, we need to think about the way our work has contributed to the valorization of white Western medicine as an unalloyed good and do a better job of resisting the great doctor or nurse narrative.

We can also be more explicit about our need to address social justice in our work from an ethical point of view. It is not simply presentism to take seriously the way that the history of health care and its racist underpinnings have created disparities. In fact, people are dying because of that history and its continuation in the present, and it is our responsibility as historians to be vocal about that. We also need to take an honest accounting of our position within our own work, assess how much we benefit from writing histories based on other people's suffering, and find ways to do less harm as historians.

NOTES

1. Some examples include Rana A. Hogarth, *Medicalizing Blackness: Making Racial Difference in the Atlantic World* (Chapel Hill: University of North Carolina Press, 2017); Deirdre Cooper Owens, *Medical Bondage: Race, Gender and the Origins of American Gynecology* (Athens: University of Georgia Press, 2017); Christopher D. E. Willoughby, *Masters of Health: Racial Science and Slavery in U.S. Medical School* (Chapel Hill: University of North Carolina Press, 2022); Martin Summers, *Madness in the City of Magnificent Intentions: A History of Race and Mental Illness in the Nation's Capital* (New York: Oxford University Press, 2019); Gabriel N. Mendes, *Under the Strain of Color: Harlem's Lafargue Clinic and the Promise of an Antiracist Psychiatry* (Ithaca, NY: Cornell University Press, 2021); and Sharla Fett, *Working Cures: Healing, Health and Power on Southern Slave Plantations* (Chapel Hill: University of North Carolina Press, 2002).

2. For summaries of these various movements, see Joan Wallach Scott, *On the Judgment of History* (New York: Columbia University Press, 2020); Catherine Hall, "Doing Reparatory History: Bringing 'Race' and Slavery Home," *Race and Class* 60, no. 1 (2018): 3–21; Ana Lucia Araujo, *Reparations for Slavery and the Slave Trade: A Transnational and Comparative History* (London: Bloomsbury, 2017); Frederico Lenzerini, ed., *Reparations for Indigenous People: International and Comparative Perspectives* (London: Oxford University Press, 2007); David Scott, "Preface: Debt, Redress," *Small Axe* 18, no. 1 (2014): vii–x; and Hilary McD. Beckles, *Britain's Black Debt: Reparations for Caribbean Slavery and Native Genocide* (Kingston: University of the West Indies Press, 2013).

3. David Scott, "Preface: A Reparatory History of the Present," *Small Axe* 21, no. 1 (March 2017): vii–x; David Scott, "Debt, Redress"; Joan Scott, *On the Judgment*; David Scott, "Preface: Evil beyond Repair," *Small Axe* 22, no. 1 (March 2018): vi–x; Hall, "Doing Reparatory History"; Karl Figlio, *Remembering as Reparation: Psychoanalysis and Historical Memory* (London: Palgrave Macmillan, 2017); Stephen Best and Saidiya Hartman, "Fugitive Justice," *Representations* 92, no. 1 (2005): 1–15; Saidiya Hartman, "Venus in Two Acts," *Small Axe* 12, no. 1 (June 2008): 1–14.

4. David Scott, "Reparatory History," viii.

5. Scott, x. See also David Scott, "Evil beyond Repair"; and Hall, "Doing Reparatory History."

6. David Scott, "Debt, Redress," viii.

7. Scott, viii.

8. Best and Hartman, "Fugitive Justice."

9. Best and Hartman.

10. "APA Apologizes for Its Support of Racism in Psychiatry," American Psychiatric Association, January 18, 2021, https://www.psychiatry.org /newsroom/news-releases/apa-apologizes-for-its-support-of-racism-in -psychiatry.

11. Jonathan Sadowsky and Kylie M. Smith, "Reflections on the Use of Patient Records: Privacy, Ethics and Reparations in the History of Psychiatry," *Journal of the History of the Behavioral Sciences* 60, no. 1 (Winter 2024): e22260; Jonathan Sadowsky, *The Empire of Depression: A New History* (London: Wiley, 2020).

12. Khalil Gibran Muhammad, *The Condemnation of Blackness: Race, Crime and the Making of Modern Urban America* (Cambridge, MA: Harvard University Press, 2010); Talitha L. LeFlouria, *Chained in Silence: Black Women and Convict Labor in the New South* (Chapel Hill: University of North Carolina Press, 2015); Sarah Haley, *No Mercy Here: Gender, Punishment, and the Making of Jim Crow Modernity* (Chapel Hill: University of North Carolina Press, 2016).

13. Hartman, "Venus in Two Acts," 10.

14. Best and Hartman, "Fugitive Justice"; Hartman, "Venus in Two Acts"; Saidiya Hartman, *Scenes of Subjection: Terror, Slavery, and Self-Making in Nineteenth-Century America* (New York: Oxford University Press, 1997).

15. Achille Mbembe, "The Power of the Archive and Its Limits," in *Refiguring the Archive*, ed. Carolyn Hamilton et al. (Dordrecht: Kluwer Academic, 2002), 19–26.

16. See Kylie M. Smith, "Jim Crow in the Asylum: Psychiatry and Civil Rights in the American South," Emory Center for Digital Scholarship, accessed November 14, 2024, http://jimcrowintheasylum.com.

17. David Scott, "Evil beyond Repair."

18. Willoughby, *Masters of Health*; Hogarth, *Medicalizing Blackness*; Brianna Theobald, *Reproduction on the Reservation: Pregnancy, Childbirth, and Colonialism in the Long Twentieth Century* (Chapel Hill: University of North Carolina Press, 2019).

19. See "Conference: Reckoning with Race and Racism in Academic Medicine," Department of the History of Medicine, Johns Hopkins School of Medicine, 2022, https://hopkinshistoryofmedicine.org/conf-reckoning -with-racism-med/.

20. Christina Sharpe, *In the Wake: On Blackness and Being* (Durham, NC: Duke University Press, 2016).

CHAPTER FOUR

Accessibility

NICOLE LEE SCHROEDER

In 2020, the world was disabled by the SARS-CoV-2 pandemic. We shifted to remote modalities and collectively prioritized the sanctity of life over the status quo. Friends, family, colleagues, and students navigated a disabled world alongside me. For the first time in my life I had remote access to research, teaching, and networking. I hoped that we would retain access gains, especially those fostering diversity, equity, and inclusivity.[1] As the years passed and public outcry lessened, though, we lost them all. People demanded a "return to normal" in 2021, remote accessibility faded in 2022, and mainstream rhetoric claims we are now "post-pandemic." For high-risk people, though, the pandemic has never ended. With each pullback, inaccessibility seeps its way back into academia.

Within months of the national shutdowns, leaders in higher education decided we were *done* with the pandemic. Conferences returned to in-person modalities and schools rescinded mask mandates. The rhetoric does not match reality. Central excess death estimates total more than twenty-eight million lives lost, and millions more have developed long COVID.[2] Roughly one in five cases results in disability.[3] That estimate is likely to increase as we learn more over time. A single infection can compromise your immune system, making you susceptible to developing a disability with each infection.[4] While academic institutions have ample resources to enforce mitigations, emotion has won out over reason. The mitigations advocated by some are dismissed as excessive, and those who advocate for them are being forced out of academic programs and research communities.

In this essay I consider access pullbacks and the demand for a "return to normal" during the ongoing COVID-19 pandemic. I argue that disabled academics shoulder excessive labor, pushback, and aggression when we demand accessibility. Mainstream academic values uphold ableist barriers that span all professional activities. As you read this essay, I ask you to consider the following: Is it ethical to work in the medical humanities while engaging in ableist practices? You will note that I use terms like *we* and *our* in reference to *my* disabled community. If your identity politics do not align with disability, I hope this offers a chance to reflect. Long have I (and other marginalized scholars) felt absent from academia's collective pronouns.

When nondisabled scholars needed access in 2020, everyone pulled together to build it.[5] Once people felt safe returning to in-person conferences, however, organizations shut down accessibility discussions and returned to an inaccessible status quo. In the past years, I have experienced accessibility pullbacks at the American Historical Association, the Society for Historians of the Early American Republic, the American Association for the History of Medicine, and many other organizations.[6] Inaccessibility is the norm *everywhere*. Access is always rationed. While disabled academics have pointed out accessibility issues in academia for decades, little has changed. Our careers are plagued by a long string of accessibility negotiations and accommodation denials. It is especially egregious, however, to witness inaccessibility in the medical humanities because our research so often centers disability. Medicine is all about how societies try to prevent, eradicate, define, or cure disability. Our field is fueled by the topic of disability, but we are rarely asked to consider our complicity in academic ableism.

Here I offer one anecdote on access labor and accommodation denials. As you read, please consider the potential harms that disabled scholars in these situations face. You might consider the following: How does it feel to do access labor? What happens when you are denied accommodations? What happens to your employment, financial security, or medical care (as, in the United States, care is dependent on employment) when you are barred from doing your job? This is not about one conference; this is about the systemic harms of academic ableism.

In 2021, a lineup of disabled historians (me included) applied to present at the 2022 American Association for the History of Medicine's national conference. As a trio of early-career scholars, we were excited by our panel's initial acceptance. Our proposal examined embodied experiences of disability as historical praxis. As a panel of high-risk scholars in Japan, the United Kingdom, and the United States, we took hope in the acceptance email, which noted a possibility for hybrid access as an accommodation. Per instruction, we immediately applied for such access. Months passed and the program launched. We noticed a series of virtual panels, but ours was not listed. Anxious as the conference date approached, we reached out. We spoke with high-risk scholars on other panels as well, but no one had heard back.

Eventually, we were told we could submit a prerecorded panel with the possibility of a Q&A. The panel would not be accessible to remote attendees. We inquired about transparency and equity. Was this a true accommodation, which would grant an equal experience to those in person? What would we gain from sharing our research without a Q&A? Behind the scenes, my copanelists and I communicated across four time zones, three continents, and divergent access needs. We compiled resources, made recommendations, and cited academic societies that have long held hybrid conferences. In the end, we were told that it was impossible to provide hybrid accommodations to everyone who had requested them.

We listened to a series of defenses regarding accessibility. It was too costly, too time consuming, or too complicated. But hybridity was possible for some panels. Programming evidenced five virtual synchronous talks, hosted by (presumably) nondisabled scholars. We questioned how these decisions were made and how access was rationed. Yet other panels were carried out in a hybrid manner because in-person panelists advocated to Zoom virtual panelists in. As a panel of four disabled and high-risk scholars, none of us were able to build access through grassroots organizing. Access and accessibility seemed to be markers of privilege. Accommodations did not mean the same thing to us that they did to organizers.

We then asked about virtual attendance, which opened up more questions about disability research. Out of 106 panel abstracts, only eight mentioned the term *disability* or *disabled*. None were made accessible to virtual attendees. High-risk people would not be privy to these conversations unless they risked in-person attendance. Disability warranted study, but disabled scholars did not warrant access. We asked why a select few panels were offered to virtual attendees, and we were told hybridity was impossible at scale. We were also informed that some presenters were unwilling to present in hybrid modalities due to worries about plagiarism. Yet these worries were seemingly absent during the 2020 virtual conference.

To be frank, these excuses for inaccessibility were nothing new. Organizers frequently claim that accessibility cannot be provided due to the required costs, labor, time, and knowledge. Disabled people are expected to fix these issues—to find funding, educate organizers, and secure services. One of my fellow panelists, Mark Bookman, engaged in this labor. Mark opened up communications, offered to consult on future access issues, and mediated correspondence. Sadly, Mark passed away in 2022. He will never benefit from the access he strove to build.

This is not just about a conference; this is about the loss of disabled expertise in the health humanities. When marginalized scholars are barred from the field or devalued within the field, the history of medicine falls flat. Earlier in this volume, Barron Lerner explains that shifts in *who* was doing medical and health history in the 1970s and 1980s prompted an explosion of diverse research. Feminist scholars, social historians, and activist-oriented historians shifted the field from "great man" history to more diverse narratives. Nevertheless, gaps remained, especially with regard to race and racism, "perhaps reflecting the relative absence of historians of color in the field."[7] Embodied experience breeds praxis, and positionality guides research innovation.

The mere existence of marginalized scholars in the field does not lead to inherent transformations. Later in this volume, Ayah Nuriddin argues that recent events, such as the murder of George Floyd, have prompted more meaningful interrogations of race and racism in medical history. Yet "Black historians in various subfields have been raising these questions for a long time, and their concerns often have not been treated as serious or relevant to the historical profession."[8] Marginalized scholars need representation in the academy, but they also need power and privilege to enact change in our community spaces. Embodied experience should be valued as a professional tool capable of challenging the status quo.

What would it look like if we collectively considered our complicity in academic ableism? Beth Linker warns, "Medical historians lack the narrative tools and language to provide an analysis of how the category of disability has operated in the past."[9] Few of us are able to access coursework in critical disability studies or emancipatory research practices, but I believe these approaches can revolutionize the history of medicine. Perhaps if we start with what we do know (our profession), we can better understand how institutions (such as academia or Western medicine) uphold ableism.

What would it look like if disabled communities could reclaim their histories and track the harms enacted on them? At present, exclusionary spaces preserve power imbalances and disabled people rarely get to conduct research.[10] As Kylie Smith notes in this volume, there is reparatory power in leveraging academic privilege to center the voices of those harmed by medicine. Ideally, we would engage in this labor *"with the goal of producing knowledge in support of justice* for people with stigmatized bodies and minds."[11] We have the potential to write meaningful histories that are responsive to the needs of disabled communities, but we must hold ourselves accountable to said communities.

Academia is deeply ableist, but *academics* are increasingly disabled individuals.[12] The National Center for Education Statistics recently reported that roughly 21 percent of undergraduate students and 11 percent of graduate students are disabled. Hiring statistics estimate a mere 1.5–3.6 percent of faculty in the United States are disabled.[13] I suspect that those percentages are even lower now, as disabled scholars have been forced out of the academy due to a lack of COVID-19 protections. What are we doing to ensure that disabled people can guide research using emancipatory frameworks?[14] What are we doing to make sure that disabled students feel represented? If we are neither uplifting disabled scholars to our ranks, nor rendering our work transparent to the public, nor connecting with modern disabled communities, how is our work ethical?

If we are serious about diversity, equity, and inclusion, we need to be serious about building accessibility. Disabled academics are scarce because they are constantly shoved out of academia. The more I call for accessibility, the more I worry about my survival in academia. Disability disclosure often leads to infantilization, discrimination, and stigma.[15] Often, when disabled people request accommodations, their petitions are not handled well. We disclose to one organizer, but others are called in to discuss. Confidentiality is not upheld. Disabled students meet with accessibility centers, faculty, and deans. Disabled faculty and staff navigate accommodations through chairs, deans, administrative departments, and human resources. We often turn to unions (if we have such protection) and hire lawyers. The more people involved in the process, the greater the risk to our enrollment, employment, or advancement status. If we refuse to apply for accommodations, we are often unable to present research, network, publish, or teach. We have to choose between two very bad options.

Academic institutions regulate access through the built environment, productivity expectations, and professional ethics. Institutions of higher education "perpetuate able-bodied hegemony, figuratively and literally constructing a world that always and everywhere privileges very narrow (and ever-narrowing) conceptions of ability."[16] This leads to "compulsory able-bodiedness," which is written into tenure files, demarcated in grant applications, and inscribed on transcripts. Ideal academics travel frequently, easily conduct research, and seamlessly juggle scholarship with high teaching loads. Disabled scholars can rarely meet these demands. Instead, we function on crip time.[17] Disability distorts time and forces us to "break in our bodies and minds to new rhythms, new patterns of thinking and feeling."[18] In the academy, however, working on crip time is forbidden. Instead, "overwork, constant work, and fast work are not only the new normal; they are framed as necessary."[19] Those who fail to maintain hyperproductivity are culled from the ranks.

Aiming for these labor standards is unhealthy for anyone, but disabled academics have to navigate accessibility labor *on top of* this unhealthy workload. Crip time is "not only about a slower speed of movement but also about ableist barriers over which one has little to no control."[20] Crip time accounts for all the time we spend communicating our access needs, negotiating accommodations, and complaining upon their denial. It includes the extra time we need to enter inaccessible spaces, survive there, and recuperate. It looks like a Deaf scholar debating with organizers over who is responsible for hiring interpreters. It is circling a building to find the accessible entrance behind the dumpster. It is grieving when you are told, per usual, that organizers simply cannot make the event accessible. It is coming up with petitions, reading lists, syllabi, workshops, and educational materials that others discard immediately.

Disabled academics who cannot perform this labor (including multiply marginalized scholars who face heightened risks when requesting accessibility) must scrape by. They can try to stay afloat in a hostile workplace or quit.[21] Leaving is rarely a choice. If you have not attended enough conferences, produced enough publications, or networked according to expectations, you are forced out. How are we supposed to build careers when so much of our energy is wasted trying to build accessibility? There is no monetary award, no CV line for "spent x hours advocating for accessibility, denied all accommodations." We want to spend our energy on things we care about, like teaching or research. Instead, we burn it away fighting for accommodations.

When accommodations are denied, disabled scholars have to do even more work to try to redress harms. This, too, is full of risk. Writing this, detailing an ableist encounter, is a risk. This work is draining and emotional, because "making a complaint is never completed by a single action. . . . It is exhausting, especially given that what you complain about is already exhausting."[22] Despite my tireless labor, my advocacy has not led to hybrid conferencing at any academic organization. The message "You don't belong here" is unrelenting.

I wonder what academia would look like if we centered accessibility. What would we gain if our spaces were open to community members? What if disabled people were there—to ask questions or critique our work? While we shy from these risks, we should embrace them. The American Historical Association's "Statement on Standards of Professional Conduct" charges us to engage in critical dialogue "with the wider public" and to secure "the trust and respect . . . of the public at large."[23] What would change if we holistically embraced the communities we write about?[24]

While academics shirk these responsibilities, disabled activists are conducting research, compiling resources, and influencing policy with projects that meet our professed ethics. Alice Wong compiles oral histories through the Disability Visibility Project.[25] Mia Mingus has theorized about the history of the medical-industrial complex on her blog.[26] Disability justice advocates write public-facing histories of medicine using websites, blogs, TikTok videos, Twitter threads, and memoirs. They combine embodied experience, historical research, and community critique. The academy at large, though, refuses to cite these works.[27]

We lose so much knowledge simply because we refuse to engage with the communities we study.[28]

I want to believe that outdated arguments for inaccessibility will fall by the wayside within my lifetime, but it is hard to imagine an accessible academy. Against our best interests, academics have upheld many barriers. We widely accept that conference presenters will fly, rent hotel rooms, purchase meals, and pay for child, pet, or elder care, often on their own dime. We accept that conferences will produce carbon emissions and spread pathogens. Scholars from the Global South, caregivers, and financially precarious scholars are pressured to accept these costs or risk further marginalization. Embracing accessibility can change some of these exclusionary trends.

If we want to embrace accessibility, we must discard excuses for inaccessibility and listen to disabled community members. Accessibility is never *too much*—neither in monetary costs nor in labor costs nor in complexity. Disabled activists already know how to build access. So it is upsetting that we are told to "be the change you want to see" but are granted neither resources, nor power, nor positionality to enact change. We need meaningful allyship to enact change.

We have already harmed community members, but we need not despair. Mingus advises, "We can only practice taking accountability when we have wronged or harmed or hurt. *Practice yields the sharpest analysis.*"[29] This is not a calling *out* but rather a calling *in*. If you finish this piece feeling angry, hurt, or frustrated—if you feel you are doing your best to build access provided structural limitations—I hope that you sit with those feelings. I feel the same every time I request accommodations, perform access labor, and find myself ostracized.

How might we build a future where accessibility standards are normalized, where no one feels angry, hurt, or frustrated by access labor? We need to begin by opening up hard conversations about funding structures, privilege, and representation. If we want to use history to enact social justice for the benefit of present-day minoritized communities, we must lead with good praxis.[30] We also have to hold on to hope, curiosity, and excitement as we pave the way forward.[31] We laid the foundation for accessible practices in 2020. How can we expand on these practices? And how might the medical humanities evolve if we choose to center accessibility?

NOTES

1. Nicole Lee Schroeder, "In Defense of Remote Access," Disabled Academic Collective, July 21, 2022, https://disabledacademicco.wixsite .com/mysite/resources.

2. "Estimated Cumulative Excess Deaths during COVID-19, World," Our World in Data, accessed April 14, 2024, https://ourworldindata.org /excess-mortality-covid.

3. "Long COVID Household Pulse Survey, 2022–2024," National Center for Health Statistics, Centers for Disease Control and Prevention, last reviewed October 3, 2024, https://www.cdc.gov/nchs/covid19/pulse /long-covid.htm.

4. Fei Gao, et al., "Robust T Cell Responses to Pfizer/BioNTech Vaccine Compared to Infection and Evidence of Attenuated Peripheral CD8+ T Cell Responses due to COVID-19," *Immunity* 56, no. 4 (2023): 864–878; Hannah E. Davis, et al., "Long COVID: Major Findings, Mechanisms and Recommendations," *Nature Reviews Microbiology* 21 (2023): 133–146.

5. Cait Kirby and Ada Hubrig, "COVID-19 and Remote Work: Access for Whom?," Disabled Academic Collective, June 8, 2022, https:// disabledacademicco.wixsite.com/mysite/post/covid-19-and-remote-work -access-for-whom.

6. Shira Lurie and Nicole Lee Schroeder, "COVID Conferences: Vulnerable Scholars Needn't Apply," *Inside Higher Ed*, October 5, 2022, https://www.insidehighered.com/views/2022/10/06/ending-remote -conference-options-exclusionary-opinion.

7. See Barron Lerner's essay in this volume.

8. See Ayah Nuriddin's essay in this volume.

9. Beth Linker, "On the Borderland of Medical and Disability History: A Survey of the Fields," *Bulletin of the History of Medicine* 87, no. 4 (2013): 535.

10. Sharon L. Snyder and David T. Mitchell, *Cultural Locations of Disability* (Chicago: University of Chicago Press, 2006), 198.

11. Julie Avril Munich, "Enabling Whom? Critical Disability Studies Now," *Emergent Critical Analytics for Alternative Humanities* 5, no. 1 (Spring 2016), https://doi.org/10.25158/L5.1.9 (emphasis in the original).

12. Nicole Brown and Jennifer Leigh, "Ableism in Academia: Where Are the Disabled and Ill Academics?," *Disability and Society* 33, no. 6 (2018): 985–989.

13. For overview, see "Students with Disabilities in Higher Education" *Postsecondary National Policy Institute* under "download factsheet" accessed 12/27/2024, https://pnpi.org/factsheets/students-with-disabilities-in-higher -education/. For full datasets, see U.S. Department of Education, National Center for Education Statistics, *National Postsecondary Student Aid Study: 2020 Undergraduate Students* accessed 12/27/2024, https://nces.ed.gov /datalab/powerstats/157-national-postsecondary-student-aid-study-2020

-undergraduate-students/percentage-distribution; U.S. Department of Education, National Center for Education Statistics, *National Postsecondary Student Aid Study: 2020 Graduate Students* accessed 12/27/2024, https://nces.ed.gov/datalab/powerstats/158-national-postsecondary-student-aid-study-2020-graduate-students/percentage-distribution; Joseph Gravely, "The Neglected Demographic: Faculty Members with Disabilities," *Chronicle of Higher Education*, June 27, 2017, https://www.chronicle.com/article/the-neglected-demographic-faculty-members-with-disabilities/.

14. There is some evidence of this in the United Kingdom. See Colin Barnes, "An Ethical Agenda in Disability Research: Rhetoric or Reality?," in *The Handbook of Social Research Ethics*, ed. D. M. Mertens and P. E. Ginsberg (London: Sage, 2008), 458–473.

15. Stephanie Kerschbaum, "On Rhetorical Agency and Disclosing Disability in Academic Writing," *Rhetoric Review* 33, no. 1 (2014): 57.

16. Robert McRuer, *Crip Theory: Cultural Signs of Queerness and Disability* (New York: New York University Press, 2006), 151.

17. Margaret Price, *Crip Spacetime: Access, Failure, and Accountability in Academic Life* (Durham, NC: Duke University Press, 2024).

18. Ellen Samuels, "Six Ways of Looking at Crip Time," *Disability Studies Quarterly* 37, no. 3 (2017), https://dsq-sds.org/article/view/5824/4684.

19. Travis Chi Wing Lau, "Slowness, Disability, and Academic Productivity: The Need to Rethink Academic Culture," in *Disability at the University: A Disabled Students' Manifesto*, ed. Christopher McMaster and Benjamin Whitburn (Bristol, UK: Peter Lang, 2019), 11–12.

20. Alison Kafer, *Feminist, Queer, Crip* (Bloomington: Indiana University Press, 2013), 26.

21. Sara Ahmed, *Complaint!* (Chapel Hill, NC: Duke University Press, 2021), 136.

22. Ahmed, 5.

23. "Statement on Standards of Professional Conduct," American Historical Association, January 7, 2023, https://www.historians.org/jobs-and-professional-development/statements-standards-and-guidelines-of-the-discipline/statement-on-standards-of-professional-conduct.

24. Kylie Smith notes that open-access publishing has reparatory power. How else can we enable access? See Kylie Smith's essay in this volume.

25. Alice Wong, Disability Visibility Project, accessed April 14, 2024, https://disabilityvisibilityproject.com/.

26. Mia Mingus, "Medical Industrial Complex Visual," *Leaving Evidence* (blog), February 6, 2015, https://leavingevidence.wordpress.com/2015/02/06/medical-industrial-complex-visual/.

27. See Courtney Thompson's essay in this volume.

28. What can we gain from disabled methods? See Mara Mills and Rebecca Sanchez, eds., *Crip Authorship: Disability as Method* (New York: New York University Press, 2023).

29. Mia Mingus, "Dreaming Accountability," *Leaving Evidence* (blog), May 5, 2019, https://leavingevidence.wordpress.com/2019/05/05/dreaming -accountability-dreaming-a-returning-to-ourselves-and-each-other/.

30. See Kylie Smith's essay in this volume.

31. For more on how to foster a growth mindset, see Cait S. Kirby, "Growth Mindset," under Learner-Facing Resources, personal webpage, accessed April 14, 2024, https://caitkirby.com/resources.html.

CHAPTER FIVE

Advocacy and Activism

JESS DILLARD-WRIGHT

As I scrawled another tally on the glass door of the intensive care unit door at 0330 one morning many years ago, something clicked. The work I was doing dragging a young person back from the absolute brink was not *done*. It would not be done at 0700 when shift change rolled around, either. Desperately unstable, this person required care that left no space or time for nursing the electronic health record (EHR). And while the EHR would not die without my ministrations, it would keep me hours beyond my shift to dutifully translate monitor-generated vital signs, hurried hash marks, harried notes, and scribbled orders into the bite-size chunks the EHR could digest. This was a watershed moment for me. I could not unsee the ways that nursing work is captured, codified, and controlled according to the demands and priorities of the necropolitical economies of health care under capitalism. In short order, this led me to other realities that I could no longer unsee. These realities were not new. But the hustle and grind of shift work in a hospital setting is busy and exhausting, perhaps by design. Too busy to think. Too busy to theorize. Too busy to organize. As I became a more expert, experienced nurse, the cracks and creaks and crevices of an unjust system became clearer and clearer. Surely, there was more to nursing and health care and hospitals, I thought, than *this*.

I left my clinical practice in pursuit of understanding in a robust way how things got the way they are. Where, I wondered, was nursing? Were we always so complicit? Why was it so difficult to communicate these questions to my colleagues, who were generally uninterested in the "whys" behind practice realities? And when nurses were in positions of power, why did they become handmaids of the bureaucracy?

So I went looking. I was already in a doctoral program studying knowledge and power in nursing and care work. Casting about for a community and narrative led me to a nurse activism think tank, where I would learn about CASSANDRA Radical Feminist Nurses Network for the first time.[1] Connecting with other nurse activists across the country provided a respite from the neoliberal grind of both health care and nursing academia. A space to think and connect. To reflect on where nursing and health care had been, to imagine where it could go. To ask, Is this version of nursing and health care really what we want?

This chapter takes up questions of activism and history through the lens of nursing in the context of ethics. In health care, ethics is often taken to mean bioethical principles or occasionally some version of care ethics focused on care encounters and correct actions. History offers up a capacious perspective for thinking about how we might conceive of ethics as a space of world-building where the thick materialities of historical inquiry give way to alternative imaginaries in which we might live, work, be, and nurse.[2] For me, a white, fat, queer, and genderqueer nurse, midwife, activist, and historian, this is a deeply personal and situated project, attuned by time and experience, echoing Richard McKay's process of becoming ethical described in his contribution to this volume. My social location in this story matters: the ethical stakes feel quite high in pushing back against the pervasively white, ableist, and cisheteronormative demands of nursing as a discipline, a discipline itself shaped by health care and society more broadly. In light of this, pushing back against the received history that nursing scaffolds its professional identity atop means recognizing that this is part of a larger political project, evoking the insights on positionality that Barron Lerner presents in this volume. The task of making space for alternative histories is tied to ethics because the histories we embrace reflect our values and build particular worlds.[3]

The nursing imaginary tangles complex realities and dreams, fraught in its legacy of engagement with advocacy and activism. In particular, I concern myself with nursing and its connections to feminism and gay liberation. This itself is an activist intervention in nursing history. Here, I will take the example of CASSANDRA Radical Feminist Nurses Network (CASS), a radical cultural feminist nursing collective, and examine their activist stance to think through how nursing has and has not engaged in political activism. This can unsettle the accepted history of nursing, making space for alternative readings of nursing's present and alternative visions of nursing's future, suggesting an ethics of possibility. This has implications both for those who aspire to become nurses and for those who will one day require nursing care. In this, CASS is a case study for how activist history can create ontological space for folks who might otherwise be imperceptible.

Locating and Creating Activist Histories

Nursing often refuses its history, particularly when it casts unflattering light on the Most Trusted Profession. Most nurses leave nursing school, enter practice, do research without a sense of nursing's history. History is deprioritized in nursing as a product of tangled discourses of anti-intellectualism and neoliberalism.[4] What little history is included in nursing education revolves around a set of white woman worthies like Florence Nightingale, a reality that serves professional identity formation.[5] While all histories are partial, the kind of histories prioritized by institutions like the American Nurses Association (ANA) and disciplines like nursing construct a ready-to-wear imaginary that props up a rigid normative ideology for nursing, who is a nurse.[6] Structured around the usual suspects of nursing history, the unassailable character and the aura of unimpeachable trustworthiness of such figures gives rise to bad readings, faded and worn simulacra presented and reinforced as singular, static, enduring, and universal. There are several reasons for this that are banal and tragic, a reality from which I will spare you here. Instead, I would like to offer up a counternarrative, making space for a queer history of and for nursing, a history that can give us different insight into the present order of things. In the section that follows, I introduce CASS, outlining some of the stakes for radical feminist and lesbian nurses in the mid- to late twentieth century. I then make the case for a queer history of nursing and conclude by attending to the ethical implications of activist history for nursing.

CASSANDRA Radical Feminist Nurses Network

Converging at the hotel bar clad in denim, flannel button-downs, and blazers, a contingent of radical feminist nurses decided that enough was enough.[7] Gathered in Washington, DC, in June 1982 for the ANA Annual Convention, these women, which included Peggy Chinn and

Elizabeth Berrey, found themselves in a city aflame, ignited in protest as the last ember of the fight for the Equal Rights Amendment (ERA) faded.[8] As feminist organizations regrouped and conservative groups rejoiced, the convention unfolded without much fanfare for the end of the ERA, a constitutional amendment designed to secure federal protections for gender equality first proposed in 1923.[9] Despite the overwhelmingly female composition of its membership, the ANA was late to support the ERA, only signing on support efforts in the mid-1970s.[10] Although an emergency resolution in support of the ERA was passed by delegates at the convention, this gesture was ultimately too little, too late, both for the amendment and for the activists who organized CASS.[11]

Active from 1982 until around 1991, CASS is unique in its radical orientation and overt identification as a radical organization for *nurses*. The formation of CASS resulted from founding members' frustrations with ANA and its avoidance of women's liberation. CASS understood ANA's slow response to the ERA as part of a protracted refusal to engage in feminist struggles, and CASS aimed to take a different path. Explicitly woman-identified, CASS was committed to "end[ing] the oppression of women in all aspects of nursing and health care," asserting "we believe that oppression of women is fundamental to all oppressions and affects all women."[12] Embracing a separatism that celebrated womanhood, CASS followed in the tradition of the Radicalesbians, embracing the "primacy of women relating to women, of women creating a new consciousness of and with each other," as a source of power for women and nurses alike.[13] This aligned CASS with lesbian separatist factions of women's liberation, like that of the legendary Furies Collective. Sharon Deevey, seasoned feminist activist and public health nurse, was a founding member of the Furies Collective along with Charlotte Bunch, Joan Biren, and Rita Mae Brown. Deevey was a central figure in CASS's Cleveland cluster. This link, as well as the group's connections to Mary Daly, Adrienne Rich, and the Susan B. Anthony Memorial Unrest Home, situated CASS squarely in the thick of lesbian separatism, even while CASS did not mandate political lesbianism from its members.[14]

Being out in nursing was dangerous, even as many lesbians populated senior leadership in the academy and profession. This was apparent when CASS founders Peggy Chinn and Charlene Eldridge Wheeler dared to submit a manuscript on feminism and nursing, which was met with friction: "In a subdued whisper . . . [they said] that while they viewed the article favorably, they could not publish it with the word lesbian in it."[15] The paper would go on to be published in 1985, including the word *lesbian*; no editors were harmed in the publication of such a scandalous word.[16] A small victory, the publication of the word *lesbian* in a nursing journal takes on added significance in the context of the witch hunts that plagued nursing leadership in the decades before CASS's formation. This meant that, even in the 1980s, CASS represented a threat because it would not stay in the closet, even if it was not all the way out of it either. Without policing sexuality explicitly, CASS endeavored to include *all women* who wished to affiliate while also prioritizing the concerns and considerations of *lesbian* nurses.[17] In service to this, CASS did not admit men, which was a source of contention for some members. It also did not shy away from lesbian identification, which created conflict when straight nurses were confronted with the assumptions of their own heteronormativity.[18] These two related issues—whether men could be members and the nuances of woman-identified sexuality—consumed much of the collective's time and energy, and they were never really settled before CASS unraveled. Letters suggest that the loose affiliation of national leadership was coming apart, a product of decentralized leadership, acrimonious succession planning, and the worsening illness of a key leader, and the organization petered out in 1991.

Toward a Queer History of Nursing

This book is about ethics in thinking, doing, and writing history, and so far I have not directly taken on the question of ethics. Yet in building a queer history of and for nursing, I am conjuring possibilities rooted in embodied, material realities, carving out historical homes for queer nurses who, like me and like the members of CASS, chafe at nursing's current state of being. These realities present nurses, historians, and nurse-historians with a choice: to shore up the status quo and remain complicit in the violence of organized forgetting—as cultural critic Henry Giroux might say, propagating a "disimagination machine" that consolidates power, knowledge, and capital in the neoliberal landscape of health care—or to resist.[19] This organized forgetting is an ongoing, active, and incomplete project in nursing as we have struggled and continue to struggle to assert a cohesive and professional identity. The respectability politics of nursing enforced and enforces a white cisheteronormativity that serves the ends of power, legitimacy, and capital. This partially explains the grind I experienced as so alienating as a clinical nurse; it also allows us to challenge forgetting as something passive or benign and highlights the political and ethical qualities of the histories we choose.

It does not have to be this way. Following Thomas Foth, Jette Lange, and Kylie Smith, we could instead think about history as articulating what is possible in the present: "To write history is basically to show that which-is has not always been like this. How we experience and understand our world depends on the meaning we attribute to things and how we interpret the context and the world we live in."[20] History can be transformative. It can transform the way we think about the present, shifting the terrain of possibility that unfolds as past, present, and future meet. This points to the ontological and ethical implications of history as a constructive project, which is where CASS comes in, at least for me.

The histories that nursing as a practice discipline uses to constitute itself in the present *matter*, reflecting the values and visions of what we choose to know and what we choose to forget. This gestures at the ethical and ontological imperatives for recognizing and claiming dissident histories for and of nursing. One aspect is reparatory, drawing on Kylie Smith's essay earlier in this volume: in spite of their ongoing invisibility, LGBTQ+ nurses have always been part of nursing and continue to be part of nursing. In many cases, they led nursing: as deans, as nurse administrators, as ANA presidents, as International Council of Nurses leaders. Mending this historical rent gives way to richer complexity for nursing, developing narratives and stories that reflect people who might dream of becoming nurses but who otherwise may not see themselves represented in the discipline. This has further implications for the folks for whom nurses care, who deserve to see themselves reflected in their care, making space to imagine what is possible by grabbing ahold of what has already been.

Activism operates in different dimensions in thinking about CASS and nursing's queer history. First, this is a *history of activism*. CASS and other lesser-known organizations like the Gay Nurses Alliance, NursesNOW, the National Black Nurses Association, and the National Association of Hispanic Nurses formed in response to the failures of ANA to push the discipline and its professional bodies to do more, do better. These organizations were linked to both liberal and radical factions of contemporaneous social movements. Leaders of the Gay Nurses Alliance, the National Black Nurses Association, and the National Association of Hispanic Nurses collaborated to push ANA from without and within to embody the truth that "the liberation of any persons from inequities contributes to the freedom of all persons" and pass the Resolution on Sexual Life-Style and Human Rights, which was adopted in 1978.[21]

This is also an *activist history* in terms of its intervention in challenging the monolithic received history that nursing instrumentalizes in service to the discipline. Because much history of nursing has been written by nurses engaged in a project of discipline building and professionalization, it has been a sanitized and sanitary history that shears off complexity and nuance in favor of respectability, an effort to leverage the power afforded by proximity to medicine. With this has come white, middle-class, cisheteronormative expectations of feminine virtue structured by patriarchy. We could choose something more expansive, more generous, more complicated.

I am not suggesting an either-or proposition, because things are always partial. The partial shift accomplished by white feminist interventions in nursing's historiography is a case in point, attending to some important issues and making a case for thinking about how gender shapes the meaning afforded to nursing work and the patriarchal "hospital family."[22] But this same shift replicated the white supremacy of earlier nursing histories, reserving critiques of power and oppression as the purview of *white* nurses directed toward patriarchy. CASS failed to embrace the intersectional paths even as the members of CASS built something defiant. This is an important lesson: you can reproduce extant power structures or challenge them. Sometimes, even when we *think* we are challenging them, we are not, or not entirely. We see this with early efforts in women's history and nursing histories and white feminisms, including CASS, where the desires and priorities of highly educated white women were understood as universal.

Conclusion

History as a tool takes on a specific meaning and function in the context of professions. Dominant histories of nursing, deployed to "apolitical" political ends, suggest that nursing is professional, clean, pure, caring, respectable, white, straight, maternal, and able. Activist history might suggest instead that this set of received ideas is a distortion, a tidy, romanticized story used to inculcate folks into a particular imaginary, public and professional. This technique of the discipline elides the heterogeneity of nurses and nursing experiences in order to legitimize and professionalize. This is a partial possibility among many, each as rooted in fact and truth and reality as the last. This leads me to a question: Why craft a queer history for and of nursing? To tell the stories—not to sensationalize. Not to glorify. Not to be able to say "gotcha." Tell the stories because they can change the world. Tell the stories because they are the stories that kids need to hear. Tell the stories because telling them repairs harm. Tell the stories because they make care possible. Tell the stories so folks who might aspire to become nurses recognize themselves. Tell the stories to find ourselves. Tell the stories because alternative views of the past can shift our understanding of the present, launching alternative paths for the future. This is world-building.[23]

NOTES

1. Peggy Chinn, "2018 Activism Think Tank Posts," *NurseManifest* (blog), 2018, https://nursemanifest.com/category/2018-activism-think-tank-posts/.

2. Jessica Dillard-Wright, "Telling a Different Story: Historiography, Ethics, and Possibility for Nursing," *Nursing Philosophy* 24, no. 3 (2023): e12444.

3. Henry A. Giroux, *The Violence of Organized Forgetting: Thinking beyond America's Disimagination Machine* (San Francisco: City Lights, 2014).

4. Thomas Foth, Jette Lange, and Kylie Smith, "Nursing History as Philosophy—towards a Critical History of Nursing," *Nursing Philosophy* 19, no. 3 (2018): 4.

5. Jennifer Woo, "Decolonize the History of Nursing by Magnifying the Contributions of Nurses of Colour," *Nursing Philosophy* 24, no. 2 (2023): 1.

6. Jamie Smith et al., "The Vitruvian Nurse and Burnout: New Materialist Approaches to Impossible Ideals," *Nursing Inquiry* 31, no. 1 (2023): e12538.

7. Peggy Chinn and Elizabeth Berrey, "Cassandra: Lesbian Nonpresence in Nursing," *Sinister Wisdom* 92 (Spring 2014): 52.

8. Marjorie Hunter, "Leaders Concede Loss on Equal Rights," *New York Times*, June 25, 1982, https://www.nytimes.com/1982/06/25/us/leaders-concede-loss-on-equal-rights.html.

9. Joanne Omang, "NOW Concedes Defeat in ERA Battle, but Vows to Make Strong Comeback," *Washington Post*, June 25, 1982, https://www.washingtonpost.com/archive/politics/1982/06/25/now-concedes-defeat-in-era-battle-but-vows-to-make-strong-comeback/01436ca6-af91-4b45-b386-0e66d883e1af/; Barbara Ehrenreich, Notebook Including Notes on

Feminization of Poverty; Sex Seminar; Citizen's Party, Spring 1982, 42–60, Barbara Ehrenreich Papers, 1922–2007, MC 565, Box 24, Folder 15, Schlesinger Library, Radcliffe Institute, Harvard University, Cambridge, MA.

10. "Historical Review," American Nurses Association, 2016, https://www.nursingworld.org/globalassets/docs/ana/historical-review2016.pdf.

11. Juanita Hunter, "Historical & Resolution; Series II; File 38," Juanita Hunter, RN and NYSNA Papers (1973–1990), January 1, 1984, Monroe Fordham Regional History Center, Archives & Special Collections Department, E. H. Butler Library, SUNY Buffalo State, https://digitalcommons.buffalostate.edu/jhunter-papers/136.

12. "Purposes," *Cassandra Radical Feminist Nurses Newsjournal* 2, no. 1 (January 1984): 3.

13. Radicalesbians, "The Woman-Identified Woman," 1970, 4, Atlanta Lesbian Feminist Alliance, Duke University Digital Repository, https://idn.duke.edu/ark:/87924/r3gx1t.

14. Peggy Chinn, Oral Herstory of Peggy Chinn, interview by Elizabeth Berrey and S. C. Warren (transcriptionist), February 2016, 38, SSC-MS-00669, Accession # 2019-S-0052, Box 1, Old Lesbian Oral Herstory Project, Sophia Smith Collection, Smith College Special Collections, Northampton, MA.

15. Chinn and Berrey, "Cassandra," 54.

16. Peggy Chinn and Charlene Eldridge Wheeler, "Feminism and Nursing," *Nursing Outlook* 33, no. 2 (April 1985): 74–77.

17. "Report of the 1985 Cassandra Continental Gathering," *Cassandra Radical Feminist Nurses Newsjournal* 3, no. 3 (September 1985): 6.

18. Peggy Chinn to Judith Carr, July 8, 1987, Cassandra Radical Feminist Nurses Network Collection, Box 1, Folder 13, Spec.201116. Cassandra, Medical Heritage Center, John A. Prior Health Sciences Library, Ohio State University, Columbus; "Report of the 1985."

19. Giroux, *Violence of Organized Forgetting*; Henry A. Giroux, "The Violence of Organized Forgetting," Truthout, July 22, 2013, https://truthout.org/articles/the-violence-of-organized-forgetting/.

20. Foth, Lange, and Smith, "Nursing History as Philosophy," 3.

21. Hunter, "Historical & Resolution," 16, 1978 ANA House Delegation Resolution on Sexual Life-Style and Human Rights.

22. Barbara Melosh, *The Physician's Hand: Work Culture and Conflict in American Nursing* (Philadelphia: Temple University Press, 1982); Susan Reverby, *Ordered to Care* (Cambridge: Cambridge University Press, 1987); Jo Ann Ashley, *Hospitals, Paternalism and the Role of the Nurse* (New York: Teachers College Press, 1976).

23. Donna Haraway, *Staying with the Trouble: Making Kin in the Chthulucene* (Durham, NC: Duke University Press, 2016), 12.

PART II

Archives and Museums

CHAPTER SIX

Decolonizing Archives and Museums

SHELLEY ANGELIE SAGGAR

Contestations over how to memorialize the colonial past have defined cultural debates in recent years, with archives and museums often at the center of these sometimes heated discussions. Although the call to confront colonial inheritances is not a new phenomenon, it is specifically *decolonization* that has become a catchall term in recent years, used to push for an array of political, material, and epistemic demands. Ranging from calls to reexamine problematic language in metadata and public displays to more material interrogations of the exploitative funding structures and sponsors that underpin the cultural heritage industry, the term has become ubiquitous in institutions over the course of the past decade. This demand to recognize how colonialism continues to shape our material and ideological present shows no signs of slowing. On the contrary, despite a more recent critical turn toward probing the utility of terms like *decolonization* in spaces—such as archives and museums—that were founded as a core part of the colonial project, activities such as community coproduction and active processes of return have largely become standard practice across a range of institutions.

Yet within these debates certain institutions have attracted considerably more attention than others. National collections, or those with ethnographic or world cultures concentrations, for example, have garnered public criticism, while archives and museums with scientific, medical, and natural history focuses have tended to evade such prominent scrutiny.[1] In this chapter I analyze this imbalance through a discussion of recent developments at Wellcome Collection, an independent museum in London that takes health and history—albeit in an expansive sense—as its primary area of interest. Reviewing the public and political response to the museum's decision to close its permanent exhibition *Medicine Man* in November 2022, I examine why science and medicine collections are still largely set apart from broader engagements with decolonization in heritage institutions. I then turn to focus on a case study of an item that was originally displayed and subsequently reinterpreted in *Medicine Man* before the gallery's closure and that serves as an example of how methods drawn from across the arts and humanities might offer historians of health and medicine the tools to address these imbalances, considering how speculative textual practices can contribute to presenting more nuanced collections stories in heritage spaces.

In a 2018 article, Subhadra Das and Miranda Lowe chart the historical connection between imperial collecting and the establishment of museums conceived of as specifically scientific institutions.[2] Reviewing the entwined roots of the now-distinct disciplines of biology, genetics, anthropology, and archaeology, the authors trace the development of taxonomy as it was applied to humans as well as other animals, and the advent of so-called race science that was advanced as a result. Das and Lowe contend that museums were central to cementing (pseudoscientific) theories of racial difference and hierarchy as *scientifically sanctioned*, "legitimizing this collecting in the context of scientific thought."[3] The role of natural science and medical museums in the creation and maintenance of colonial hierarchies is, the authors argue, a central aspect of both their historical mission and their current position. For historians of health and medicine, critically unpacking this inheritance is useful in interrogating discourses of neutrality that continue to underscore how the health sciences and museums themselves are popularly conceived of in both public and scholarly imaginations.[4]

A recent example illuminates this point. In 2022, following several years of research, critique, and incremental curatorial revisions, Wellcome Collection publicly announced its intention to close the long-running exhibition *Medicine Man*.[5] The gallery, which opened in 2007, presented the story of Sir Henry Wellcome's historical collection, a broad endeavor that primarily aimed "to tell a global story of health and medicine" through a vast array of artistic, cultural, and scientific artifacts.[6] The objects in the gallery were organized by theme rather than type, but nevertheless, items—including human remains—were displayed with minimal context and through a lens that largely occluded autochthonous perspectives. Even the design of the exhibition (dark wood paneling with minimal lighting and limited interpretation) worked to create a somewhat nostalgic evocation of a gentleman's collection of curiosities.

Following the announcement of *Medicine Man*'s closure, Wellcome Collection found itself subjected to a barrage of criticism from the press, politicians, and social media commentators.[7] While some praised the museum's ability to critically reexamine its historical and continuing colonial dynamics, others were dismayingly quick to dismiss the entire initiative. These responses rejected the invitation to participate in a collaborative conversation that Wellcome Collection had extended, instead framing the debate as merely a virtue-signaling exercise carried out by vandals in a continuing culture war.

Until this point, despite heightened commentary regarding colonial inheritances in contemporary British cultural institutions, Wellcome Collection had remained largely unscathed by public criticism. I would like to suggest that this is primarily due to two factors that are important for both museologists and historians of health and medicine to consider in critical scholarship. The first is that, unlike national institutions in the United Kingdom, Wellcome Collection retains a certain independence due to its status as a private collection (although many commentators on Twitter were quick to demand that the museum be stripped of its—nonexistent—public funding). Second, and most important for those interested in critical approaches to the history of health and medicine, Wellcome Collection's thematic focus had allowed it to remain relatively unscathed in the contemporary culture wars, which have largely focused their attention on anthropological, ethnographic, and world cultures collections. Science and medicine often rely on precisely the same mythologies of objectivity and neutrality that museums also shroud themselves in, allowing them to sit out highly charged debates regarding the coloniality of their disciplinary histories.

Yet, as many scholars have shown, despite this convenient mythology, scientific disciplines, including health and medicine, are not free of bias, prejudice, and flattening assumptions regarding the validity of other epistemic models and practices.[8] Indeed, in the rush of online responses to Wellcome Collection's announcement, Twitter users were quick to rehearse binary distinctions between "scientific enlightenment" and "marginal" and "archaic methods" that illustrated precisely these prejudices.[9] These examples, albeit inflamed by the rhetorical posturing of social media's discursive styles, nevertheless demonstrate the persistence of popular conceptions of Euro-Western science and medicine as preferentially valued compared with other epistemic methods. Moreover, it is interesting to note that despite Wellcome Collection's specific focus on addressing the roster of "racist, sexist and ableist theories and language" in the *Medicine Man* gallery, many of the comments in response to the museum's announcement homed in on colonial distinctions between an imagined "us" (enlightened, scientific, rational) and a vague "they" (relativistic, religious, cultural).[10]

In countering these kinds of problematic divisions and disciplinary assumptions, however, archives and museums often find themselves rehearsing a certain set of stories. Such narratives strive to recognize the deliberately diminished representation of the colonized in history but, in so doing, adopt a tone that laments this erasure while stopping short of pointing to incisive critical methods that might offer the potential for invoking more creative interpretive methods that can account for nuance and complexity. In the United Kingdom especially, archives and museums resound in a self-reflexive chorus, reciting the various ways in which their founders *encountered*, *entangled*, and *enmeshed* the peoples of the world amid the machinations of the British empire. While this admission of imperial involvement is certainly a welcome departure from the hurried erasure of any mention of empire that has often accompanied archival and museum narratives, uncritically repeating such a meticulously crafted disclaimer serves to abstract the stories of the people represented in these collections and neglects any engagement with narratives that depart from flattening assumptions about colonial display and Indigenous agency. In grappling with this pattern, I now turn to examine an example from my own work with Wellcome Collection that aimed to complicate this default response by drawing out latent stories that were present, yet neglected, in the original *Medicine Man* display. In discussing this project, I go on to suggest the creative potential of turning to speculative and cross-disciplinary methods in recovering narratives of Indigenous agency and presence in historical medical collections in particular.

Restoring Mana, Recovering Story: The Life Mask
of Taupua Te Whanoa

In 2019, I spent a year researching sacred, secret, and otherwise culturally sensitive items in the remainder of the collection of the Sir Henry Wellcome Historical Medical Museum. Through presenting four case studies drawn from a subcollection of over 4,000 objects, I devised an approach that aimed to highlight the care, commitment, and attention such research requires over a long-term, even lifetime, basis.

One item I focused on was a plaster cast mask depicting the face of a Māori man. The mask, which had been displayed in Wellcome Collection's *Medicine Man* exhibition since its inception in 2007, contained remarkably minimal information about who the person represented was or how they came to be displayed in a gloomy gallery on the Euston Road. Instead, the mask was juxtaposed with a piece of tattooed human skin from a French sailor, gesturing toward the history of French presence in the Pacific and rehearsing a narrative of benevolent imperial *encounters* at work in the *Medicine Man* exhibition more generally.[11] But assuming a singular story of acquisition without agency ultimately contributes to a culture of curatorial work that, while well intentioned, continues to deny those represented in archives and museums the histories and futures of presence and power that have always been enacted by colonized peoples around the globe.

The Wellcome Collection plaster cast is actually a copy; the original mask was made from a living person, in contrast to those that were cast from *toi moko*, or the preserved heads of Māori.[12] The Wellcome Collection copy contains a handwritten insert recording assumed provenance, but this information had been mislabeled or misinterpreted, leading to difficulties in ascertaining the mask's precise origins. "G. Grey" (referring to Governor Sir George Grey, the original mask's owner) had become "G. Guy," and "Tauque Te Whanoa" had been recorded as the name of the man represented.

Te reo Māori, however, contains no *q* letter, a fact that suggested that the man had been misnamed in the museum's records.[13] These initial clues led me instead to a figure named Taupua Te Whanoa.[14] Following a research inquiry made to Te Papa Tongarewa (the national museum of Aotearoa / New Zealand), I confirmed that the Wellcome mask

Plaster cast, man of the Arawa tribe showing Maori tattooing.
Wellcome Collection, https://wellcomecollection.org/works/qg56wpsn
/images?id=fq5gnf97. License, Attribution 4.0 International (CC BY 4.0).

did, in fact, represent him, and both Dougal Austin (Kāti Māmoe, Kāi
Tahu, Waitaha) and subsequently Paul Tapsell (Ngāti Whakaue, Ngāti
Raukawa) generously shared more information pertaining to Te Taupua's life and the history of the cast.[15] Tapsell noted that Te Taupua was
the eldest son of Te Whanoa and a descendant of Pukaki of Ngāti

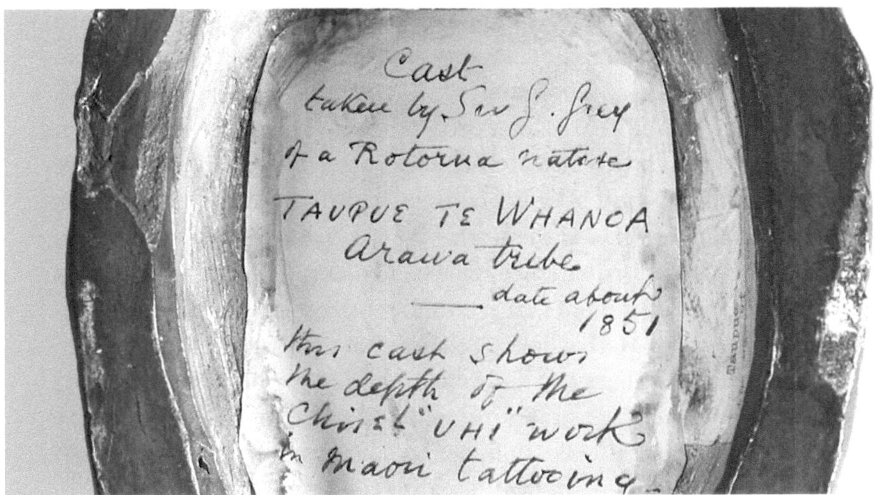

Interior view of the mask, showing handwritten provenance information. Wellcome Collection, https://wellcomecollection.org/works/ewtwpf9q /images?id=bbcapb6c. License, Attribution 4.0 International (CC BY 4.0).

Whakaue, Rotorua. Te Arawa is a confederation of *iwi* and *hapu* in the Rotorua region; thus, the "Arawa tribe" notation on the Wellcome mask is a rather clumsy way to note the man and mask's relational affiliations. Tapsell also shared that Te Taupua was a renowned carver who lived at Ohinemutu and that he met with Sir George Grey in late December 1849 at Te Ngae, or the Bay of Plenty. He was persuaded to have his tattooed face cast and expressed no issue with the plaster cast being taken, despite the discomfort of the process. It is worth emphasizing that despite these difficulties, the limited provenance we were able to glean from the handwritten labels offered significantly more information than many items in such collections, including Wellcome's. As Amanda Mahoney discusses in her contribution to this part of the volume, a lack of adequate resources, staffing, and space often results in this key information being lost, leaving researchers interested in decolonizing archives and museums at an immediate disadvantage.

Unearthing the stories of those who were brought to Britain to be exhibited as live displays to paying customers, historian Sadiah Qureshi contends that "even within the confines of evidently restrictive situations, displayed peoples created ways to maintain their agency."[16]

Te Taupua—in his hybrid status as a historical figure and also as a displayed in-gallery representation—can be understood to have enacted precisely this kind of strategic navigation in order to assert his own agency in the representational process. Thus, the information recovered through this research restores agency to Te Taupua, as we understand him, in light of his consenting to have his likeness transported to the other side of the world via his plaster-cast representation.

In 2020, I was invited to contribute to a curatorial intervention as part of the redevelopment and eventual replacement of Wellcome Collection's *Medicine Man* gallery. Although the initiative did not seek to "decolonize" the gallery as a singular ambition, the final contributions went some way toward grappling with the inheritance of Sir Henry Wellcome's historical collection and its meaning to visitors and staff interested in the history of health and medicine in the present day.

For my contribution to this project, I sought to restore Te Taupua's story to his displayed representation and address some of the elisions of imperial history that continue to color presentations of health and medicine in museum collections. Previously, the interpretive label attached to the display gave an abstract and technical account of the process of applying *ta moko* (tattooing). This original label text frames the plaster cast as a typified example, disregarding any information specific to the man represented. Moreover, the interpretation strongly emphasizes masculinity, recalling imperial mythologies of Māori hypermasculinity and their status as a so-called warrior race.[17] Although sometimes subtle, these elisions and assumptions in the original label contributed to an overarching narrative of sensationalized difference—a feature that defined the curatorial interpretation and organization of the original *Medicine Man* gallery.

My intervention sought instead to restore three crucial aspects of Te Taupua's story: first, his name, which was absent in the original display. In her essay in this volume, Adria L. Imada considers the ethics of naming historical agents in research, arguing for a practice of historical scholarship that carefully attends to the myriad textures of life, data, and colonial erasure that shape who is and is not named in historical work. To me, it felt important to refuse the museum's aestheticization of the cast by naming the man represented. I hoped to signal, as Imada suggests, the subjectivity and personhood of Taupua Te Whanoa. I also set out, second, to restore his relational affiliations, and third, to draw out a speculative narrative of agency that reinterpreted his presence in the *Medicine Man* gallery. My revised interpretation was displayed alongside the original label and a commissioned response in te reo Māori from Te Horowai, also a descendant of Pukaki of Ngāti Whakau/Te Arawa, which provides additional detail and perspective on Te Taupua's story.[18]

Plaster cast showing

Maori tattooing

The first step in applying a Maori tattoo or 'Moko' was to sketch the pattern with charcoal. The pattern was then carved into the skin with a chisel to create grooves. Soot was then rubbed into the skin to provide colouring. The process could take days or weeks depending on the individual's tolerance for pain. Every man wore an individualised pattern. Women were also tattooed, but not as elaborately as men.

New Zealand, 1851

A62970, Wellcome Collection / Science Museum Group*

Plaster cast life mask of Taupua Te Whanoa, A642970, Wellcome Collection/Science Museum Group

This is not an object. This is a person with a story.

The man represented is Taupua Te Whanoa. He has been displayed here since 2007, but until now, was unnamed. Te Taupua was a descendant of Pukaki of Ngāti Whakaue. In 1849, he consented to having a plaster-cast likeness made. This is a copy of the original.

Without a story, the mask was little more than a colonial "curio." By restoring his name, ancestry and agency, we can imagine a life for him beyond the confines of this glass case.

Te Taupua's story exceeds this display, it is up to us to recover it.

Thanks to Paul Tapsell and Dougal Austin for assisting with this research†

Te Taupua—he tohunga whakairo o Ngāti Whakaue ki Ōhinemutu

---

* Original text of display label.
† New interpretation by Shelley Angelie Saggar.

This is one of a number of casts from the original, which was procured

from Te Taupua himself on the occasion Sir George Grey visited our tribal territory in 1849.

Koineki tētehi o ngā tākai I hangā ki te tākai mataati e Te Taupua I a Sir George Grey I takahe haere ana in ngā whenua Māori I te tau 1849.

Te Taupua was the first child born in Ōhinemutu after Ngāti Whakaue captured this village from a rival tribe, from whom he also descended.

Ko Te Taupua I whānau tuatahi mai ki Ōhinemutu i muri mai i te pāhoronga o te whenua e Ngāti Whakaue mai i téétehi iwi wheinga, ā, nō taua iwi tonu.

Te Taupua is remembered for his carving expertise; he tohunga whakairo. In particular, he carved Pukaki (of Te Māori fame) in 1836 as a young man.

Nā ngā mahi whakairo a Te Taupua i whetū tārake ai. Nā Te Taupua a Pukaki i whakairo i te tau 1836 i a ia taitāheka ana.

This massive gateway honoured his great grandfather who led the Ōhinemutu battles.

He whakamaumaharatanga te waharoa nui whakaharahara nei ki tōna koroua i tū hai tātāriki I ngā pakanga i Ōhinemutu

His final masterpiece was the large elaborately carved meeting house of Ōhinemutu named Tamatekapua.

I whakairohia rākaitia te whara tupuna a Tamatekapua e Te Taupua, Ka mutu, koineki te mounga whakamutunga a Te Taupua

It was opened in 1876 just prior to Te Taupua passing away.

I tūherangia i te tau 1876 mua tata tonu atu i te matenga o Te Taupua

TE HOROWAI

DESCENDANT OF POKAKI OF NGĀTI WHAKAUE / TE ARAWA,

ROTORUA, AOTEAROA NEW ZEALAND, WHERE THE CAST MASK OF TE TAUPUA WAS ORIGINALLY MADE 172 YEARS AGO.‡

‡ New interpretation by Te Horowai.

Speculative Fictions: A Methodological Proposition

There are still aspects of Te Taupua's life and journey that remain unknown. While speculation might seem incompatible with traditional museum or historical methodologies, such methods, drawn from disciplines such as literary studies, present a compelling opportunity to craft and recover agential narratives in the colonial collections we have inherited. In her 2008 essay "Venus in Two Acts," Saidiya Hartman conceives of "critical fabulation" as a route to restore, and restory, narratives of presence and agency among those whom the historical archive has violently silenced.[19] In Hartman's case, it is important to note, these gendered figures are drawn from the archive of Atlantic enslavement—a specific context that I seek to think in conjunction *with*, rather than uncritically transpose, given the very different context that Te Taupua's display emerges from. Recovering the biographies of those who are made epistemologically unknowable through deliberate elisions and silences, this intersection of archival work and speculative writing is theorized as a way of exceeding the "fictions of history" that at once exposes the fantasy of museal and archival neutrality and imagines liberatory possibilities in the cracks and at the edges of the historical record, formally conceived.[20] In light of the renewed information we have recovered from other scholars' careful and committed understandings of Te Taupua's life and relations, I want to suggest we might speculatively reinterpret Te Taupua as a kind of nineteenth-century cosmopolitan traveler, rather than solely as an exploited displayed person constrained within a one-sided colonial power relation. In this way, we might restore Te Taupua's mana—earned prestige and power—to the mask in Wellcome's collection by using these alternative methods to tell this much fuller story.[21]

Looking to these alternative methods, historians of health and medicine might consider how such methods might offer opportunities to reconstruct historical narratives from fragmented and silenced sources. The example of Taupua Te Whanoa's plaster-cast mask in Wellcome Collection's *Medicine Man* gallery gestures to the wider issues of power, (re)presentation, and agency at work in the stories we continue to tell about the history of health and medicine in heritage institutions. As Wellcome Collection redevelops its displays, these "enforced silences" that remain embedded in how we present our disciplinary histories must be addressed through methods that center and respect autonomy and narrative control.[22] Yet as historian Coll Thrush points out in his 2016 book *Indigenous London*, Indigenous peoples' stories never need *discovering* by others.[23] Rather, currents of agency and historical presence are latent in archives, museums, and the land itself. Indigenous peoples' stories far exceed the confines of glass cases and gallery displays; it is merely the role of historians and heritage workers, Indigenous and otherwise, to restore these.

NOTES

1. Paul Daley, "The Gweagal Shield and the Fight to Change the British Museum's Attitude to Seized Artefacts," *Guardian*, September 25, 2016, https://www.theguardian.com/australia-news/2016/sep/25/the-gweagal-shield-and-the-fight-to-change-the-british-museums-attitude-to-seized-artefacts; Laura Van Broekhoven, "Committed to Change," Pitt Rivers Museum, accessed May 10, 2021, https://www.prm.ox.ac.uk/committed-to-change.

2. Subhadra Das and Miranda Lowe, "Nature Read in Black and White: Decolonial Approaches to Interpreting Natural History Collections," *Journal of Natural Science Collections* 6 (2018): 4–14.

3. Das and Lowe, 5.

4. Saroo Sharda, Aruna Dhara, and Fahad Alam, "Not Neutral: Reimagining Antiracism as a Professional Competence," *Canadian Medical Association Journal* 193, no. 3 (January 18, 2021): E101–E102; LaTanya S. Autry and Mike Murawski, "Who We Are," Museums Are Not Neutral, accessed February 17, 2022, https://www.museumsarenotneutral.com/who-we-are.

5. "Statement on the Closure of Our Medicine Man Gallery," Wellcome Collection, November 28, 2022, https://wellcomecollection.org/pages/Y4TdMBAAACMApB14.

6. Wellcome Collection, "Medicine Man," Wellcome Collection, accessed February 10, 2023, https://wellcomecollection.org/exhibitions/Weoe4SQAAKJwjcDC.

7. BBC Politics (@BBCPolitics), "Conservative MP Ben Bradley says there are a 'handful of woke nutters' in charge of museums who are too concerned about the artefacts they display, as the Wellcome Collection closes its Medicine Man exhibition after 15 years," Twitter, November 29, 2022, https://twitter.com/BBCPolitics/status/1597608140987461634.

8. See, for example, Donna Haraway, "Situated Knowledges: The Science Question in Feminism and the Privilege of Partial Perspective,"

Feminist Studies 14, no. 3 (1988): 575–599; Robin Wall Kimmerer, *Braiding Sweetgrass: Indigenous Wisdom, Scientific Knowledge and the Teachings of Plants* (London: Penguin Books, 2020); and Katherine McKittrick, *Dear Science and Other Stories* (Durham, NC: Duke University Press, 2021).

9. Stephen Bremner (@sbremner98), "@ExploreWellcome Please curl up in a ball of mindless self-loathing and disappear. Unfortunately for you the scientific enlightenment was of massive benefit to humanity. Indigenous medicine was not. Any attempt to try and relativise this is doomed to be pure propaganda and will be despised," Twitter, November 26, 2022, https://twitter.com/sbremner98/status/1596587640609980416.

10. Shatmonkey (@braidedriver2), "@ExploreWellcome They were marginalised because they're marginal and their archaic methods have no place in scientific discovery," Twitter, February 19, 2023, https://twitter.com/braidedriver2/status/1627210823191006082.

11. I use the term *encounter* here to refer to the problematic way in which imperial histories and meetings are often discussed in museums. For more on the problems with this framing, see Shelley Angelie Saggar, "Encounter," *Decolonial Dictionary* (blog), July 10, 2019, https://decolonialdictionary.wordpress.com/2019/07/10/encounter/.

12. Marc Fennell's *Stuff the British Stole* offers a general introduction to the history of *toi moko*. Marc Fennell, "Stuff the British Stole: The Headhunters," *Stuff the British Stole*, podcast, December 12, 2020, https://www.abc.net.au/radionational/programs/stuff-the-british-stole/the-headhunters/12868026.

13. I am grateful to my former colleague at the Science Museum, Margaux Wong, for initially noting this.

14. There are various spellings of Taupua Te Whanoa's name. I follow Te Papa's convention.

15. Paul Tapsell, *Pukaki: A Comet Returns* (Auckland: Reed, 2000).

16. Sadiah Qureshi, *Peoples on Parade: Exhibitions, Empire, and Anthropology in Nineteenth-Century Britain* (Chicago: University of Chicago Press, 2011), 153.

17. Clare Barker, "Warrior Genes," *MFS: Modern Fiction Studies* 66, no. 4 (2020): 755–779.

18. The first line of Te Horowai's intervention contains a misprint that was, unfortunately, published in the gallery. This should read, "Te Taupua—he tohunga whakairo o Ngāti Whakaue." I was not directly involved in the liaison with Te Horowai, but I would like to use this space to apologize for the mistake nonetheless.

19. Saidiya Hartman, "Venus in Two Acts," *Small Axe* 12, no. 2 (2008): 11.

20. Hartman, 9.

21. Anaïs Walsdorf, "Why Wellcome Closed Its Medicine Man Exhibition—and Others Should Follow Suit," Conversation, December 16, 2022, https://theconversation.com/why-wellcome-closed-its-medicine -man-exhibition-and-others-should-follow-suit-196171.

22. Coll Thrush, *Indigenous London: Native Travelers at the Heart of Empire* (New Haven, CT: Yale University Press, 2016), 6.

23. Thrush, 6.

CHAPTER SEVEN

Advising around Human Remains Collections

MELISSA GRAFE

Regularly I hear some version of the following question when leading tours into the warmly lit room full of the brains, tumors, and photographs of the people formerly under the care of renowned neurosurgeon Harvey Cushing (1869–1939): "Did the patients give consent to have their brains and tumors collected?" It is a question I wrestle with every time, torn between the historical aspects of patient consent in the past and the realities of it in the present, the ethics around the display of human remains, and the sheer wonder of the Cushing Center. And it is a question that is much larger than one person can answer.

The center is the home of the Cushing Tumor Registry, built by Cushing and his colleagues between 1902 and 1932 and encompassing nearly 750 human brains and tumors; approximately 15,000 glass-plate negatives of patient images and histology; and slides.[1] The center is a late addition to the field of medical museums, opened in 2010 in the subbasement of the Harvey Cushing / John Hay Whitney Medical Library. It was built primarily in response to larger environmental threats damaging the registry, as all the remains and photographs were formerly stored in the subbasement of the medical student dormitory, which lacked any climate control. Turner Brooks, architect for the center, consciously created a warm, inviting space for the display of over 300 jars, patient photographs, Cushing memorabilia, vitrines for book exhibitions, and twenty-two interactive discovery drawers. In its first decade, thousands of people from inside and outside Yale visited the center, including middle school classes that incorporated the collection into curriculum on neuroscience.

The center stems from a tradition of anatomical museums in medical schools and societies that stretches back to the eighteenth century in America and earlier in Europe.[2] While these museums rarely exist anymore, as specimens fell into disuse and new technologies arose to peer into human bodies, a few collections made transitions into medical museums. For example, the Warren Anatomical Museum in the Center for the History of Medicine, Countway Library of Medicine at Harvard University, is "one of the last surviving anatomy and pathology museum collections in the United States."[3] While the gallery is temporarily closed as part of a redesign, museum staff offer tours of and educational sessions on the collection, which dates back to 1847. The Mütter Museum of the College of Physicians of Philadelphia, originally established in 1859 for the medical education of its medical members, has been open to the public for several decades, currently attracting over 130,000 visitors annually.[4] The Maude Abbott Medical Museum at McGill University in Montreal, Quebec, Canada, was established in 2012, although its collection, similar to those of the Cushing Center, Warren Anatomical Museum, and Mütter Museum, was assembled by medical professionals in the nineteenth and early twentieth centuries.[5] Interest in public displays of human bodies also extends to circulating exhibitions like *Body Worlds* and *Bodies: The Exhibition*, which run circuits across various science museums and galleries throughout the United States and internationally.[6]

Museums are not the only pathway for access to human remains. Medical schools have all kinds of research collections containing pathological specimens, many used for research and teaching purposes only, which was the original purpose of the Cushing Tumor Registry. The Percival Bailey Brain Specimens Collection at the University of Illinois Chicago, for example, is only available online for public viewing.[7] Bailey, a neuropathologist trained in surgery by Cushing, was an instrumental part of the development of the Cushing Tumor Registry, involved in the study, classification, and care of the specimens. The Brain Collection at the University of Texas at Austin, in a similar arc to that of the Cushing Tumor Registry, was stored in a closet for a number of years. But these specimens, collected by resident pathologist Coleman de Chenar between 1952 and 1983, belonged to patients with mental disorders at the University of Texas State Mental Hospital. There was discussion about placing this brain collection on display at the

Imaging Research Center at the University of Texas, but a permanent exhibition had not materialized as of 2024.[8]

Historical human remains collections containing enslaved, colonized, Indigenous, and vulnerable individuals are increasingly controversial, particularly in light of the larger racial reckoning in the United States after the George Floyd murder.[9] In 2020 and early 2021, the Morton Cranial Collection at the University of Pennsylvania Museum of Archaeology and Anthropology, built by physician and anatomist Samuel George Morton and used for teaching and research purposes, came under scrutiny for its possession of Black remains torn from burial grounds, as well as Morton's larger racial legacy.[10] Many other museums and institutions with human remains collections began researching the origins behind their acquisition and the individuals represented in the remains. For example, Harvard University created the Steering Committee on Human Remains in University Museum Collections in 2020, which issued a broad report in September 2022.[11] Very little is said in the Harvard report about the human remains on display in museums already, other than a bulleted line on the final page: "Display will be governed by careful consideration given to the context of the acquisition of the remains and the teaching rationale for their exhibition."[12] These parameters around exhibitions, teaching, and acquisition are only part of the larger story for many human remains collections, including those of the Cushing Center.

At the heart of all the discussion on the display of humans is the vulnerable position of the people whose bodies become part of these collections. From human fetuses to those hospitalized for mental disorders, the power dynamic regularly falls in favor of the medical industry. Yale medical student Aminah Sallam identified some of the particular issues around the display in a 2019 article and case study of the Cushing Center. She notes, among other recommendations related to privacy and curation, that "in the cases where consent was not (or cannot be) obtained, a discussion of the historical acquisition of specimens on display should be included in the exhibit."[13] Sallam refers to the International Council of Museums' code of ethics, which requires that human remains "be presented with great tact and respect for the feelings of human dignity held by all peoples."[14] She also points out the jarring contrast between the center's display and the trust and confidentiality at the heart of medical training, including a culture of patient protections reinforced by HIPAA (the 1996 Health Insurance Portability and Accountability Act). In light of the renewed advocacy around the ethical stewardship of human remains collections and issues of privacy and consent that Sallam describes, how should policy on display and the remains in the Cushing Center be crafted? And as the center's collection is not HIPAA protected, how do we balance the training that medical students receive, which emphasizes anonymizing the patient, and the individual stories that historians often tell in the course of their research and writing to humanize what has become a medical specimen?

Display and the Rise of the Advisory Board

Policies on the display of human remains in museums are often left to individual institutions, which are struggling to formulate how to address issues around exhibition on top of the inherent ethical problems related to these collections. Display relates not only to how objects are mounted and shown in museum spaces but also to how materials are interpreted to various audiences. The Cushing Center's display was built, in part, on medical student Christopher Wahl's interpretation of the history of the collection, which formed the basis of his 1996 medical school thesis.[15] Wahl created a temporary exhibition that was displayed in part at the 1996 American Association of Neurological Surgeons annual meeting and, later, outside the Cushing/Whitney Medical Library at Yale.[16] The 2005 publication of a new biography on Harvey Cushing by Michael Bliss brought additional material for the interpretation of the center, as did a 2007 book by Aaron Cohen-Gadol and Dennis Spencer, with patient profiles drawn from the Cushing Tumor Registry.[17] Ultimately, despite the hundreds of brains and tumors quietly floating in jars, as well as over a dozen patient photographs, the story of the Cushing Center focused on Harvey Cushing and his major role in the development of modern neurosurgery, with very little attention paid to the individuals represented in the collection. Most labels were adapted from Wahl's exhibition and the scholarship on Cushing, with minimal labeling throughout the center. The discovery drawers and main exhibition case featured Cushing objects, the vitrines contained books from Cushing's library, and the poster display held oversize images of Cushing's life, his books, and patient montages. The scripts for guided tours reinforced this veneration of Cushing, as initial audiences for the center were mainly drawn from the medical community.

With a very limited budget and only a part-time coordinator for the center, changes in display and interpretation happened slowly and somewhat organically over time. Staff members and graduate students selected and curated new book exhibitions in the vitrines several times in the first decade, with the most recent exhibition focused on the history of the development of the registry and center for its tenth anniversary in 2020. As audiences broadened beyond the medical community, the coordinator selected and added new posters and provided educational materials and books for school groups and others who wanted to dive more deeply into neuroscience or Cushing's life. In one of the biggest changes, the discovery drawers were completely redone just before the center's tenth anniversary. New interpretation moved the focus away, in part, from Cushing, to encompass more patient stories and a larger acknowledgment of the often-hidden labor around the creation of the registry and the people in Cushing's orbit. The tour scripts partially changed as well, in response to growing audiences beyond the medical community who came to the center for various reasons, such as classroom teaching, research, interest in art or photography, or the novelty of the space. Our guides, including myself, continually adapted the tours based on questions from audience members, bringing in new information as we learned more from our community. We also wrestled with the morbid allure of the space, informally nicknamed "the Brain Room" in the Yale community, and worked to reinforce the connections to the people represented by the brains and photographs.[18]

In late 2019, in light of ongoing questions from museum patrons on consent and privacy and broader discussions around Sallam's article, I recognized the larger need for guidance on the center's curation and display beyond the changes just described. The center suffers from confusion around who has actual ownership, authority, and control of the space and the collection, which has a major impact on ethical considerations. While library staff take care of the collection and offer some interpretation, the Department of Neurosurgery funds the coordinator, and the center represents a larger cultural space in the School of Medicine. I strongly felt the library needed to bring in larger community participation and discussion of the challenges around ethical display, starting with baby steps. I approached the director of the

Cushing/Whitney Medical Library to discuss the creation of a volunteer advisory board. Advisory boards are a common tool in the museum and library field, bringing outside viewpoints into display and collection decisions, as well as helping to build connections to larger communities. Unlike focus groups, friends' groups, task forces, or ethical review boards, advisory boards often provide a sustained sounding board over a longer course of time, usually with some rotating membership. Surprisingly, advisory boards for the care of human remains collections, or their display, do not appear to be common, although the field is small.

While the COVID-19 pandemic put the formation of the board on hold, the director and I began crafting a charge for the board and had some initial discussions on membership. We searched for other medical museum models and eventually offered the following charge for the eventual review of the advisory board. The charge clearly centers the ethical and academic issues, whatever may arise, as a major priority: "Provide support and advice on ethical and academic issues related to the Cushing Center, help draft mission and vision statements for the Center, advise and assist in the development of new programs, and help identify best practice standards. Board members may also serve as ambassadors on behalf of the Cushing Center, providing a connection to and ongoing exchange of information more broadly."

The selection of members was and continues to be iterative. For the initial board formation, we considered the expertise within the Yale community, although we recognized the connections to the larger New Haven and patient communities, who are audiences of the center. We sought a mix of views representing medicine, ethics, museums, and history and looked at people engaged with the center in various ways, some closely connected, some critical, and others who rarely came into the space but had a deep understanding of the history of human remains or the practices around human remains collections. During our initial meeting in fall 2022, advisory board members identified other possible members for consideration, and the addition of medical students is a likely next step. The center's advisory board also began working on the larger issues surrounding the display of the collection and directed center and library staff toward the crafting of interpretation of consent. In later meetings, the process of crafting position statements on the display of human remains and educational mission highlighted the need for a larger interpretive shift in the exhibition itself and triggered a larger debate on how to center personhood in the remains left of the individuals represented in the registry. The center's advisory board provided rich and sometimes contradictory viewpoints concerning the ethics of the display, something that the library will have to negotiate as new policies and larger display changes are considered. The ethics are messy and confusing, but the addition of new voices is helping to clarify some of the tangled threads the library needs to carefully consider going forward.

Conclusion

An advisory board cannot solve all the institution's issues, and the institution may act on the advice in unexpected ways as policy is formed. But it is the richness of the experiences of advisory board members, and the viewpoints they bring to issues facing the institution, that provides institutional staff with pathways toward policies that can alter the shape of museum interpretation and collection. While the Cushing Center advisory board is still in its infancy, the Morton Cranial Collection at the University of Pennsylvania Museum of Archaeology and Anthropology, mentioned earlier in this chapter, provides an example of how advisory board members shape policy decided by institution. Abdul-Aliy Muhammad, a community organizer on the museum's advisory committee on repatriating the remains, objected to the museum making any decisions on the burial of thirteen skulls of Black Philadelphians in the collection, "criticiz[ing] what they called the museum's 'rushed' decision-making and [saying] descendant communities should decide what happens to the remains."[19] This objection is causing the museum to think more carefully about its role in determining the final resting place of the remains and who has power to make those decisions. As legal, ethical, and cultural forces continue to shape the ultimate fate of human remains collections, my hope is that the Cushing Center advisory board will help steer the center as it navigates new landscapes in display and care.

NOTES

1. "About the Cushing Center," Cushing Center at Yale's Harvey Cushing / John Hay Whitney Medical Library, accessed November 22, 2022, https://library.medicine.yale.edu/cushingcenter/about.

2. For a larger discussion on the history of anatomical and medical museums, see Michael Sappol, *A Traffic of Dead Bodies: Anatomy and Embodied Social Identity in Nineteenth-Century America* (Princeton, NJ: Princeton University Press, 2004); Ann Fabian, *The Skull Collectors: Race, Science, and America's Unburied Dead* (Chicago: University of Chicago Press, 2010); Shauna Devine, *Learning from the Wounded: The Civil War and the Rise of American Medical Science* (Chapel Hill: University of North Carolina Press, 2014); Samuel J. M. M. Alberti, *Morbid Curiosities: Medical Museums in Nineteenth-Century Britain* (Oxford: Oxford University Press, 2011); and Samuel J. Redman, *Bone Rooms: From Scientific Racism to Human Prehistory in Museums* (Cambridge, MA: Harvard University Press, 2016).

3. "Warren Anatomical Museum Collection," Center for the History of Medicine, Harvard Countway Library, accessed November 15, 2024, https://countway.harvard.edu/center-history-medicine/collections-research -access/warren-anatomical-museum-collection.

4. "About," Mütter Museum, College of Physicians of Philadelphia, accessed November 26, 2022, https://muttermuseum.org/about/overview.

5. "Introduction," Maude Abbott Medical Museum, McGill University, accessed November 26, 2022, https://www.mcgill.ca/medicalmuseum /introduction.

6. There was controversy around the origins of some of the bodies plastinated for display, at least when these types of exhibitions became popular. Gunther von Hagens's *Body Worlds* exhibitions and knockoffs like *Bodies: The Exhibition* and *Real Bodies: The Exhibition* have all been accused of sourcing bodies from executed Chinese political prisoners or Russian homeless people, indigent hospital patients, and prisoners. See Neda Ulaby, "Origins of Exhibited Cadavers Questioned," *All Things Considered*, NPR, August 11, 2006, https://www.npr.org/2006/08/11 /5637687/origins-of-exhibited-cadavers-questioned; and Colin Perkel, "'Bodies Revealed' Exhibit May Be Using Executed Chinese Prisoners, Says Rights Group," CBC, September 6, 2014, https://www.cbc.ca/news /canada/bodies-revealed-exhibit-may-be-using-executed-chinese-prisoners -says-rights-group-1.2757908.

7. "About This Collection," Percival Bailey Brain Specimens Collection, University of Illinois Chicago, accessed November 26, 2022, https:// collections.carli.illinois.edu/digital/collection/uic_neuro (link no longer active).

8. Jessica Riley Holmes, "Brain Collection at the University of Texas at Austin," Handbook of Texas Online, last updated January 12, 2021, https://www.tshaonline.org/handbook/entries/brain-collection-at-the -university-of-texas-at-austin.

9. This is not the first time the larger political climate affected human remains collections. In 2005, medical anthropologist Lynn M. Morgan found herself at the center of a controversy around abortion and a collection of human fetuses at Mount Holyoke College in South Hadley, Massachu-setts. Lynn Marie Morgan, "The Rise and Demise of a Collection of Human

Fetuses at Mount Holyoke College," *Perspectives in Biology and Medicine* 49, no. 3 (2006): 435–451. No enslaved or Indigenous remains are in the center's collection, based on a preliminary review of existing records for many of the remains in the center. More research needs to be completed.

10. Samuel J. Redman, "Bodies of Knowledge: Philadelphia and the Dark History of Collecting Human Remains," *Perspectives on History*, September 15, 2022, https://www.historians.org/perspectives-article /bodies-of-knowledge-philadelphia-and-the-dark-history-of-collecting -human-remains-october-2022/; Stephan Salisbury, "Origins of Skulls Held by Penn Hard to Untangle: Potter's Fields, Which Lie Beneath the Campus, Were Dug Up. Human Remains Were Removed for Study," *Philadelphia Daily News*, February 16, 2021, https://www.proquest.com /newspapers/origins-skulls-held-penn-hard-untangle/docview/2489520497 /se-2.

11. Steering Committee on Human Remains in University Museum Collections, *Report of the Steering Committee on Human Remains in University Museum Collections* (Harvard University, Fall 2022), https:// provost.harvard.edu/files/provost/files/harvard_university-_human _remains_report_fall_2022.pdf.

12. Steering Committee on Human Remains in University Museum Collections, 31. As of February 2024, as this essay was in review, the Smithsonian Institution, in response to newspaper articles focused on human remains in its collection, issued a report from its Human Remains Task Force that stated, among other principles. that "human remains should not be displayed by the Smithsonian (in exhibition, print or online) unless done so with the documented and informed consent of the deceased or, in appropriate circumstances, their descendants or descendant communities." Smithsonian Institution, "Human Remains Task Force Report to the Secretary," January 10, 2024, 3, https://www.si.edu /sites/default/files/about/human-remains-task-force-report.pdf.

13. Aminah Sallam, "The Ethics of Using Human Remains in Medical Exhibitions: A Case Study of the Cushing Center," *Yale Journal of Biology and Medicine* 92, no. 4 (December 20, 2019): 768.

14. International Council of Museums, *ICOM Code of Ethics for Museums* (Paris: International Council of Museums, 2017), 25, https:// icom.museum/wp-content/uploads/2018/07/ICOM-code-En-web.pdf.

15. Christopher Wahl, "The Harvey Cushing Brain Tumor Registry: Changing Scientific and Philosophic Paradigms and the Study and Preservation of Archives" (MD thesis, Yale University, 1996), https:// archive.org/details/39002086341949.med.yale.edu.

16. Parts of Wahl's American Association of Neurological Surgeons exhibition was available online as of 2022: Christopher Wahl, "Gone but Never Forgotten: Renaissance of the Harvey Cushing Brain Tumor Regis-try," American Association of Neurological Surgeons, accessed November 30, 2022, https://www.aans.org/cybermuseum/tumorregistryhall/wahl .html, site no longer available.

17. See Michael Bliss, *Harvey Cushing: A Life in Surgery* (New York: Oxford University Press, 2005); and Aaron A. Cohen-Gadol and Dennis D. Spencer, *The Legacy of Harvey Cushing: Profiles of Patient Care* (New York: Thieme, 2007). Spencer, the Harvey and Kate Cushing Professor and chair of the Department of Neurosurgery at Yale University School of Medicine at the time, was instrumental in the resurrection of the collection and the creation of the Cushing Center. For more on the development of the center, see Dennis Spencer, "Harvey Cushing's Oxalis: The Cushing Center at Yale University," *Interdisciplinary Science Reviews* 38, no. 3 (2013): 210–221.

18. In a different case study, new interpretation of three fingers of a murder victim, which have been the focus of morbid curiosity for visitors at the Wood County Historical Center and Museum in Ohio, is discussed in Rebecca Mancuso, "The Finger Saga: One Museum's Quest to Turn the Macabre into the Meaningful," *Public Historian* 40, no. 2 (2018): 23–42.

19. Remy Tumin, "Penn Museum Grapples with How to Honor Skulls," *New York Times*, August 11, 2022, https://www.proquest.com /newspapers/penn-museum-grapples-with-how-honor-skulls/docview /2700467260/se-2. As of February 2024, the Penn Museum had interred the remains of nineteen Black Philadelphians, following a legal challenge from Muhammad and the Black Philadelphians Descendant Community Group over the museum's plans for the remains. See Peter Crimmins, "Penn Museum Has Laid to Rest 19 Skulls from Its Morton Cranial Collection," *Philadelphia Tribune*, February 2, 2024, https://www.proquest .com/newspapers/penn-museum-has-laid-rest-19-skulls-morton/docview /2927866252/se-2.

CHAPTER EIGHT

Identifying the Human in Human Remains Collections

AISLING SHALVEY

When I began my PhD study in Strasbourg, I was one of a group of academics intending to uncover the remains of Nazi medical research, including possible human remains.[1] My PhD was a component of the historical commission of investigation into the former Nazi university the Reichsuniversität Straßburg (1941–1944). While I was mentored by a team of established academics and learned a lot from them, I was initially surprised that there was no instruction to influence how we would deal with this collection. This lack of clear signposting on what to do next spoke to the complexity of the situation.

The city of Strasbourg, on the border of France and Germany, had been part of the German Empire until the end of World War I (1918), when it became French. Retrieval of this city in 1940 was important to the Nazis to symbolize their presence in the West. In turn, they established a university to epitomize Nazi science and research. The university was opened with the aim to "dethrone the Sorbonne" (one of the leading French universities) and replace it with a university that produced science grounded in Nazi ideology; this included unethical medical research on marginalized groups.[2] This new Reichsuniversität Straßburg's unethical research materials included the skeleton collection of the anatomist August Hirt, compiled by "ordering" eighty-seven Jewish people from Auschwitz concentration camp, the plan being to kill them and use their skeletons to "prove" racial difference.[3] The guiding principle in establishing the commission of investigation into the Reichsuniversität Straßburg medical faculty was that there is "no smoke without a fire"—the presence of three war criminals in one faculty could indicate further unethical research that had yet to be discovered.[4]

In 1945 the university became French again, with a new name and new staff, and while the physical location of the faculty had not changed between Nazi occupation and French return, everything else about the faculty had.

This historical commission was preceded by the discovery of human remains in the legal medicine department in 2015.[5] These remains included a stomach, containing only potato peelings, which came from a Jewish victim of the Holocaust in the concentration camp Natzweiler-Struthof. This proved definitively the malnutrition and deprivation experienced by the victims. These remains had been collected and classed as evidence for the war crimes trial in the French city of Metz but had remained in the collection following the autopsy. A collective burial and memorial were created in the Jewish cemetery in Strasbourg, and a street was renamed for the first identified victim, Menachem Tafel.[6] The University of Strasbourg hoped that by creating a historical commission of investigation, possible further remains could be identified.

I knew that finding these remains, if there were any, was one of the main goals of the commission and my PhD. Alongside this was the goal of understanding the structure of the former medical faculty and ascertaining who else may have been using this institutional background to conduct unethical experiments. On the afternoon of October 1, 2018, I climbed to the attic of the anatomical institute with a colleague to find paraffin blocks for her research on syphilis treatment in the 1920s. I discovered paper slips attached to glass jars, albeit very dusty and neglected for seventy years, and realized that these were remaining samples from the Nazi era.[7] I did not expect, though, that once these remains were found, there would be no rulebook on how to deal with the discovery. I sketched a rudimentary map identifying where things were and how extensive the collection was. These remains were not entirely forgotten; it was known that remains were in the attic from a former public collection, but as they had been removed from display and never investigated, the history department assumed that they were innocuous medical teaching collections with no links to the Nazis. This discovery illustrates one of the main points of this essay: that merely removing collections from public view and keeping them in storage does not actively engage with the ethical problems of the collections.

With institutional support, I was given clearance to move specimens, carefully clean the paper slips, and begin identification where possible.

This involved cataloging the attic where they were kept to try to ascertain any clear groupings that might indicate a research collection. The slides were cleaned of dust with a fine brush until writing could be seen, sometimes typed but often handwritten in pencil. When some identification was visible—a name, a year, or a dissection number—this was noted. I then compared the pathology records, however incomplete, with the remaining patient files. Based on this, I built up a clinical picture of what had happened to the victim and whether their death and the preparation of a sample were a result of medical research or a result of a routine operation. While time consuming, and often unsuccessful, the process was quite an emotional one. I was the first person to speak their names in eighty years, and it was unknown who the last person to handle the specimens was. The unknown weighed heavily on my mind, particularly the possibility that after all this work we may never know why this collection was compiled, or by whom. That I could not do a literature review to figure out more about the collection was distressing and took me completely out of my comfort zone. That was what I had been trained to do; that was where I felt supported and comforted by scholars who went before me. In the absence of that familiar structure, I was forced to confront what I was seeing and ask myself what my own thoughts were, unmediated through previous work. I looked at how others dealt with similar collections, and their questions on ethics, then began to formulate some of my own. I managed to definitively identify 134 specimens, but not all could be traced with a patient file, given routine deaccession of hospital records over the years. When I could not achieve complete identification, those people I felt I had "failed" stuck in my mind. I wrote a piece about those who went unidentified in the *Polyphony*, and I found it resonated with others—trying to pretend I was the ideal objective historian was impossible when face-to-face with a baby who had gone unburied for seventy years.[8]

When the time came for writing the report on the findings, I dealt with further internal conflict concerning the presentation of physical remains and photographs of them.[9] In order to prove to the commission funding body that findings had emerged, photographs would be needed, and it was their responsibility to then decide what to do with these remains: retain them in collection, bury them, cremate them, or return them to family members. I compiled specimens for a press release and collected photographs from the day I found the specimens for the report.[10] I felt an internal conflict between publicizing photographs of human remains to increase awareness that such remains still exist in university collections and displaying such photographs as a continuation of the medical gaze, thus encouraging the viewer to see them as "other." In the process of identification, we ascribe the name and family history to the individual, removing them from the category of unnamed, unidentified "other" and encouraging the viewer to see that this could be one of their family members. While such discussions might seem polarizing in trying to decide what is ethically right or wrong, it speaks to a moment of transition and conversation where it is acknowledged that there is no singular right or wrong. In the case of Strasbourg, it became evident to us all in the commission that these remains required further analysis, having been definitively identified as not a result of unethical research or victims of the Holocaust. These specimens came from individuals who had routine surgeries or who had natural deaths and were autopsied under normal circumstances, although their remains were retained without any indication their family were aware of the situation. It has thus far been decided by the University of Strasbourg, in consultation with the historical commission, that these specimens will not be buried, as they appear to be a result of clinical practice and teaching rather than experimentation.

This collection of human remains was constructed only because of the Nazi ideology that humans were not equal. It was retained for so long only because of the dismissive attitude that it was compiled by someone else and so should be addressed by someone else. But avoiding unpleasant realities does nothing to address systemic injustice and only further allows such injustice to remain in place and unquestioned. Racist research was carried out in the context of "ordinary" medical research, and the only way we can deal with human remains collections is to identify provenance where we can, rather than surrender to a knee-jerk reaction to conceal that which we find uncomfortable about our past. By publishing these findings, it is hoped that we contribute to the conversation on this spectrum of ethical responses and reactions, on dealing with the individual, and on making decisions based on provenance rather than general assumptions, which in turn may be culturally insensitive. Ultimately, best practice evolves over time with ethical considerations, and resistance to critique or to change actively hinders the process of doing ethical history. Discussions of ethics and of origins need to include archives and museum collections, but this discussion should not stop there.

In some cases, burial itself can become fraught with ethical debate. This is the case with the Max Planck Society collection concerning brain specimens from Nazi research.[11] I am currently the project coordinator for this group, but the search for human remains in the collection, identification of these remains, and the burial of them has been an ongoing task since the 1980s. Götz Aly, a German historian, stated that the remains in teaching collections in Germany came from victims of Nazi biomedical research. In many cases, the family remained completely unaware that the victim's brain was retained for research, having been given the body for burial. From 1939 to 1945, Nazi scientists collaborated to acquire more material for their experiments and for scientific analysis through the murder of prisoners of war, psychiatric patients, and other groups.[12] While this experimentation and murder were known in the aftermath of the war, and widely publicized through the Nuremberg doctors' trials that later led to the Nuremberg Code, this focused exclusively on the treatment of live patients and not the specimens that remained in medical collections across Europe.[13] Not only were entire brains, other organs, and microscopic slides retained in collections, they were still in use as teaching material for medical students and research material for scientists. By the 1970s, medical students, journalists, historians, and social scientists had publicized their concern with the enduring legacy of medical harm that was not addressed by medical schools.[14] This was something of an open secret, with Julius Hallervorden, an acclaimed scientist, having collected brains of people who were killed for research purposes—because he did not kill them himself, he felt himself completely innocent.[15] Hallervorden stated to Leo Alexander, the American psychologist tasked with examining the perpetrators before their trial, "Of course I accepted the brains. It really wasn't my concern where they came from and how they were brought to me."[16] He also noted earlier that he was purely concerned with research, and when asked how many brains he could process, he said, "The more [brains] the better."[17]

In 1989 the University of Tübingen in West Germany had identified all of the anatomical specimens used in postwar medical teaching that had come from victims of Nazi injustice. The commission decided to remove the specimens from the collection and follow up with a collective burial.[18] This burial at Gräberfeld X was meant to be a final ethical and symbolic act, and to function as an example for other collections that remained part of medical schools.[19] The commission that was founded to oversee this regarded the anonymous burial as an example. This included both those suspected to be victims and those confirmed as victims, and the commission considered it, as historian Paul Weindling states, to be drawing a line under the past and thus achieving closure with a moral course of action.[20] The Max Planck Society, which was also a repository for many human remains used for scientific research and teaching, conducted a similar burial in 1990 in the Waldfriedhof in Munich for all specimens from the Nazi period from 1933 to 1945 (the society then was called the Kaiser Wilhelm Society). The burial was collective and anonymous, with the reason for not establishing provenance noted as being that it would take too long to identify victims, which numbered in the thousands. As Amanda Mahoney has noted in this volume, provenance can be extremely problematic and difficult to decipher, but an attempt can be made, and many names and origins can be uncovered. The collection was far more extensive than previously thought, involving other institutions across Germany. The current project that I work on is aimed at redressing the confusion caused by the anonymous burial without provenance research at the Max Planck Society in 1990.[21] This can be done by identifying which specimens were buried, identifying which specimens remain in the collection, and crucially, constructing biographies of the victims of brain research. Many specimens that were supposed to be buried have remained in the collection, while others were only discovered many years after the burial. This project shows the need for transparency and individual identification of specimens.

It is important to note that student protests, public understanding of the retention of remains, and the breaking of this collective silence had only taken place for ten years or so by the time these remains were buried.[22] At that time, most historians considered a more fitting burial place to be enough. Certain procedures, such as sterilization or hanging for so-called crimes, were still considered to be highly personal and private, and many family members did not wish their relatives to be explicitly named.[23] As the years pass and we get further from World War II and the atrocities it involved, people are more willing to speak, knowing that there was never a valid reason for their relatives to be targeted. In the 1990s, the most ethical, moral, and sensible method of bringing justice to the victims seemed to be a collective anonymous burial. That this decision might later be looked on as too hasty, ill-advised, or incomplete did not dissuade researchers from doing the right and ethical thing in that moment. We can, and should, be empowered to do what is ethical in the present moment, even if this decision needs to be reevaluated in later years.

As time went on, these collective anonymous burials began to include memorial stones that listed the names of the victims and thus created a place for commemoration and mourning, acknowledging that each victim was not just one in several thousand but an individual person with a family that should be made aware of their resting place and what happened to them. Kylie Smith's chapter in this volume goes into further depth on the issue of reparatory history, arguing that it is pertinent not only to more recent human remains collections, as in this case for Holocaust victims, but to all collections where an imbalance of power is evident. Today, identification is in the best interests of the victim and can in part undo some of the anonymization ascribed to them by the perpetrators when referring to them in medical journals only as numbers and not as people.[24] This identification and open naming, however, should not be considered the standard; each collection needs to be considered individually. The burial in Tübingen remains at the forefront of ethical practice, as calls to reopen the caskets and identify the victims were listened to. As a result, in 2022 the commission re-formed to investigate this.[25] The Tübingen example has highlighted that solutions are not permanent, and in fact the search for a permanent perfect solution may hinder ethical progress entirely. While multiple policy documents have been constructed to deal with this situation, notably the Vienna Declaration, such documents often do not enable ethical solutions for specific situations and cannot be relied on to answer collection-specific questions.[26] The Vienna Declaration deals with how to treat human remains discovered to be from victims of the Holocaust, but it has no advice regarding reburial, the testing of remaining samples, or retrospective identification of already buried remains. There is a general reluctance to dig up, both figuratively and literally, human remains information for provenance research, in part because the answers are often more unclear the more in depth we go.[27]

When I began to work with this entirely new collection in Strasbourg and searched for a road map on what to do, and how to do so ethically, I realized there was no such standard. Realizing there is no road map for navigating the ethical pitfalls speaks to how we can all grow in our research. Overall, it must be considered that the scope, time, and resources available for this type of work are integral to our ethical processing of these collections, and this can affect whether we view such methods as proactive ethics or reactive ethics. One of the central things I have learned is that the model of the ideal objective historian is not only a myth but in itself unethical. To remain objective and unemotional when dealing with human remains collections or the Holocaust speaks not to an ethical objectivity but to a cold callousness that fails to recognize and leave space for the humanity of victims. While institutions can present solutions that may be very temporary and transient, their temporality does not detract from the fact that we have tried to heal some of the harm. These solutions to human remains collections must make space for the natural emotional responses both of the historian and of the community that is involved. There is no one single ethical answer when it comes to human remains collections, but if there were, it would start by saying, "Do less harm."

NOTES

1. Aurelien Breeden, "A French University Confronts Medical Crimes and Its Nazi Past," *New York Times*, July 24, 2022, https://www.nytimes.com /2022/07/24/world/europe/french-university-nazi-medical-crimes.html.

2. Tania Elias, "La cérémonie inaugurale de la Reichsuniversität de Strasbourg (1941)," *Revue d'Allemagne et Des Pays de Langue Allemande* 43, no. 3 (July 2011): 341–361.

3. Hans Joachim Lang, "August Hirt and 'Extraordinary Opportunities for Cadaver Delivery' to Anatomical Institutes in National Socialism: A Murderous Change in Paradigm," *Annals of Anatomy* 195 (2013): 373–380.

4. "University of Strasbourg Investigates History of Medicine during Nazi Era," Arolsen Archives, accessed November 15, 2024, https://arolsen -archives.org/en/news/university-of-strasbourg-investigates-history-of -medicine-during-nazi-era/.

5. Florence Rosier, "For 30 Years the University of Strasbourg Claimed That It No Longer Held the Remains of Nazi Victims, but This Was Not True," *Le Monde*, May 11, 2022, https://www.lemonde.fr/en/france/article /2022/05/11/for-30-years-the-university-of-strasbourg-claimed-that-it-no -longer-held-the-remains-of-nazi-victims-but-this-was-not-true_5983071_7 .html.

6. "Des restes d'expériences des médecins nazis," *Le Figaro*, July 20, 2015, https://www.lefigaro.fr/flash-actu/2015/07/20/97001-20150720 FILWWW00162-des-restes-d-experiences-des-medecins-nazis.php.

7. Jean-Marie Le Minor et al., *Anatomie(s) & Pathologies: Les collections morphologiques de la Faculté de Médecine Strasbourg* (Bernardswiller, France: I. D. l'Édition, 2009), 125–166.

8. Aisling Shalvey, "Naming the Invisible Patient," *Polyphony*, June 18, 2021, https://thepolyphony.org/2021/06/18/naming-the-invisible-patient-a -historical-dilemma/.

9. For further discussion on the ethical use of images in historical work, see the chapter by Michaela Clark in this collection.

10. The photographs were reproduced in the *New York Times* and *Le Monde* (previously referenced), among others; originals are available in Gabriele Moser, Aisling Shalvey, and Paul Weindling, "L'Institut de pathologie et ses collections et archives," in *La faculté de médecine de la Reichsuniversität Straßburg et l'hôpital civil sous l'annexion de fait nationale-socialiste 1940–1945: Rapport final de la Commission historique pour l'histoire de la faculté de médecine de la Reichsuniversität Straßburg*, ed. Christian Bonah, Florian Schmaltz, and Paul Weindling, (Strasbourg: Université de Strasbourg, 2022), 387–413, https://www.unistra.fr/fileadmin/upload/unistra/universite /historique/Rapport_final_Reichsuniversitat_Strassburg_corr.pdf.

11. Paul Weindling et al., "The Problematic Legacy of Victim Specimens from the Nazi Era: Identifying the Persons behind the Specimens at the Max Planck Institutes for Brain Research and of Psychiatry," *Journal of the History of the Neurosciences* 32, no. 2 (2023): 218–239.

12. Weindling et al.

13. *Trials of War Criminals before the Nuremberg Military Tribunals under Control Council Law No. 10* (Washington, DC: US Government Printing Office, 1949), 2:181–182, https://history.nih.gov/display/history/Nuremberg %2BCode.

14. Jürgen Peiffer, "Phases in the Postwar German Reception of the 'Euthanasia Program' (1939–1945) Involving the Killing of the Mentally Disabled and Its Exploitation by Neuroscientists," *Journal of the History of the Neurosciences* 15, no. 3 (2006): 210–244.

15. Heinz Wässle, "A Collection of Brain Sections of 'Euthanasia' Victims: The Series H of Julius Hallervorden," *ENDE Endeavour* 41, no. 4 (2017): 166–175.

16. Herwig Czech, Paul Weindling, and Christiane Druml, "From Scientific Exploitation to Individual Memorialization: Evolving Attitudes towards Research on Nazi Victims' Bodies," *Bioethics* 35, no. 6 (2021): 508–517, at page 511.

17. Jurgen Peiffer and Michael Shevell, "Julius Hallervorden's wartime activities: Implications for science under dictatorship," *Pediatric Neurology* 25, no. 2, (2001): 162–165.

18. Tübingen University Abschlussbericht der Commission zur Überprüfung der Präparatesammlungen in den medizinischen Einrichtungen

der Universität Tübingen im Hinblick auf Opfer des Nationalsozialismus, MPG II. Abt., Rep. 1F, Az A-II-7a Besondere Aufgaben Hirnschnittsammlung, July 13, 1989.

19. Paul Weindling, "Hiding in Plain View: Burial and Commemoration of Children's Specimens from Wittenau in the 'Gräberfeld/Cemetery X' Tübingen, 4 and 8 July 1990," *Medizinhistorisches Journal* 56, no. 3 (2021): 219–235.

20. Weindling.

21. "Max Planck Society Concludes General Audit—Victim Research Project to Commence in June 2017," Max-Planck-Gesellschaft, May 2, 2017, https://www.mpg.de/victims-research-project.

22. Peiffer, "Phases in the Postwar."

23. Peiffer.

24. Paul Weindling, "The Need to Name: The Victims of Nazi 'Euthanasia' of the Mentally and Physically Disabled and Ill 1939–1945," in *Mass Murder of People with Disabilities and the Holocaust*, ed. Brigitte Bailer and Juliane Wetzel (Berlin: International Holocaust Remembrance Alliance and Metropol, 2019), 49–82.

25. "Uni Tübingen: Suche nach Überresten ermordeter Kinder aus NS-Zeit erfolglos," *SWR Aktuell*, March 17, 2023, https://www.swr.de /swraktuell/baden-wuerttemberg/tuebingen/uni-tuebingen-reste-aus-ns -grab-untersucht-100.html.

26. Joseph A. Polak, "'Vienna Protocol' for When Jewish or Possibly-Jewish Human Remains Are Discovered," in *How to Deal with Holocaust Era Human Remains: Recommendations Arising from a Special Symposium*, ed. William Seidelman, Lilka Elbaum, and Sabine Hildebrandt (Boston: Boston University School of Public Health, November 22, 2017), 12–19.

27. For further discussion of an individual case study on searching for provenance of a Māori mask at Wellcome Collection, see Shelley Saggar's essay in this volume.

CHAPTER NINE

Stewardship of "Challenging" History of Medicine Collections

AMANDA L. MAHONEY

As chief curator of a large history of medicine collection, I am charged with the stewardship of ethically challenging historical materials. My role requires that I maintain and interpret collections held in the public trust. Proper stewardship involves many difficult decisions, all of which have ramifications for the scholars, students, and visitors who access our collections. Often, the practicalities of stewardship, particularly of ethically complex materials, are obscured from the public, sometimes creating the impression that information is being withheld or decisions are being made without careful consideration. Quite the contrary is true; most museum, archive, and library professionals working with historical health care collections take a highly principled approach to interpreting, exhibiting, and providing access to the primary sources in their care. The problem is that these professionals have limited resources and power, and these are often insufficient to do the "right" thing regarding an artifact or create meaningful change within their institution. The lack of transparency surrounding how historical repositories make stewardship decisions adds to the sense that archivists and curators do not care about ethical practices or that the barriers to allowing access to "sensitive" materials are personal rather than institutional, legal, or the result of insufficient resources. Museums and other repositories often have incomplete information regarding their collections, insufficient resources to maintain records, and inadequate infrastructure for making their collections truly accessible to scholars and visitors alike. This lack of information and support makes ethical decision-making regarding the use and care of challenging materials in history of medicine collections particularly difficult.

In this essay, I will discuss a specific example drawn from my experiences as curator and historian at the Dittrick Medical History Center to illustrate the challenging and convoluted process of deciding what to "do" with ethically difficult elements of historical medical collections. By sharing this decision-making process, I hope to offer an alternative to all-or-nothing attitudes that persist regarding ethically difficult collections. I will also describe common challenges faced by my colleagues in museums, archives, and library special collections for researchers unfamiliar with how the sausage is made in historical repositories, to encourage empathy and collaboration.

The Dittrick, part of Case Western Reserve University in Cleveland, Ohio, is a small university museum housing artifacts and archives related to the history of medicine. Like several other medical history repositories in the United States, the Dittrick was founded as the artifact collection of a local medical association (ca. 1894) and later became affiliated with a university. The collections as well as their accompanying documentation reflect this history, as does the Dittrick's corresponding shifts in mission, curatorial goals, and audience. The artifacts themselves reflect many changes in health care over the past few centuries, such as shifting attitudes toward the patient body and the patient-physician power dynamic.

One of the most ethically fraught items in the Dittrick's collections is not an "artifact" or "object" but rather a human skull previously used by a physician as a desk ornament (a candy dish, pencil cup, or ashtray). The skull presents many stewardship challenges, including questions of how and whether it should be exhibited or interpreted, who should have access to it, and how to appropriately handle, store, and keep track of it within our existing infrastructure. Even something as simple as how to refer to the skull during internal meetings is not straightforward—it is not an artifact or object, *it* seems to further objectify the remains, and *he* or *him* feels like a slippery slope toward giving the skull a cutesy nickname.[1] And yet, we need to be able to discuss it and make practical decisions about the skull's future even as we wrestle with the broader issue of human remains held within museum collections. I often find myself framing the skull interchangeably as an artifact (e.g., when updating the collections database), biological material (e.g., when deciding the best environmental conditions for its storage), and a person (e.g., when I am examining the skull or talking about it to others). Each approach is practically useful in my role as curator, but none of these frameworks feels entirely correct or comfortable.

Thinking of the skull as an "artifact" is unfortunately necessary in order to follow best professional practice. For instance, it is unrealistic to think that our small team could appropriately care for the skull without including it in our collections database. The preservation of biological materials requires environmental conditions and collections care expertise that differ from those required by the typical history museum artifact. Because of the skull's fragility, it is currently stored in one of our most well-controlled environments alongside nonbiological artifacts.

The donor, a collector of surgical tools, purchased the skull along with a few instruments from a yard sale held by a physician upon retirement during the early 1970s. The physician had already donated his professional papers, instruments, and other items to a local history repository. The donor believed the skull was used as an ashtray and later repainted for another use. All of this information was verbally communicated between the physician and donor and later sent to the Dittrick via email—we do not have any documentation from before our exchanges with the donor. According to the donor, the physician acquired the skull "during his medical school days" (ca. 1920). There are no further details regarding its person or the physician's role (if any) in transforming the skull into a vessel.

While this knowledge of the post-1970s history of the skull is useful regarding its care and provides some information about its origin via a medical school or hospital, we know nothing about its person, their life history, or their wishes regarding their remains after death. The recent history of these remains is well documented, yet we do not know their full provenance. Provenance, data on the history of an artifact, is key to the appropriate and ethical stewardship of a historical object, as well as intrinsic to its value as a primary source for scholars.[2] Provenance information is critical to many aspects of how museums "use" physical collections, including creating exhibits that convey an item's historical context, employing it as a primary source in scholarly research, and ensuring its ethical stewardship.[3] While the concept of provenance is more closely associated with art museums, it is perhaps even more important within the context of historical museums and special collections. The educational value of a historical artifact is closely tied to the context of its creation, use, and ownership history. Detached from this context, an object loses varying degrees of its usefulness as a primary source.

The tangibility of the skull, its physical appearance and material composition, is a compelling link to the medical world of the 1910s–1920s and a visceral reminder of institutional contempt for the dead, poor, and sick at that time. I personally feel the most uncomfortable with my utilitarian approach of framing the skull as alternatively an object-artifact, biological material, or a person during physical examination and requisite handling, an integral part of collection stewardship. This is in large part due to the physical evidence that documents the initial transformation of it from (likely) the remains of a dissection cadaver into a decorative object, which feels to me like a desecration.

These human remains were intentionally dehumanized to create a vessel for display. The top of the skull has been crudely cut away to form a basket-like handle. The interior of the skull is painted black with a lacquer-like finish. A large cork was stuffed into the foramen magnum and cut at an angle, presumably to allow the skull to rest on a flat surface. Nails were used to secure the jaw to the skull on either side but no longer serve this function, leaving the jaw loose. While no markings such as initials are extant, there are multiple signs that indicate a lack of expertise with anatomical preparation. For example, cuts along the frontal and parietal bones are rough and some tissue still remains around the teeth. The small nails used to attach the lower jaw were inserted into fused joints between skull bones (sutures), which no experienced preparator of anatomical specimens would choose, as it is likely to accelerate the structural deterioration of the skull.[4]

The upper teeth are in poor condition. The teeth of the lower jaw are more or less intact, displaying signs of enamel erosion and decay that may reflect the wear and tear sustained by the person during their lifetime. The teeth of the upper jaw are all broken, some shattered down to the root, with corresponding damage to the teeth of the lower jaw. I was saddened to learn from a forensic pathologist that this damage is commonly found on historical anatomical skeletons, the result of people repeatedly manipulating the lower jaw to make the skull "talk." While there is some evidence that abscessed teeth and other infections may have contributed to the death of the skull's person, it is not possible to determine cause of death or definitively identify the circumstances that resulted in the individual's remains ending up on a dissection table.

The structural changes made to the skull go well beyond the interventions necessary for a cadaver dissection or autopsy. It was not simply "left over" from a medical procedure or educational exercise. Rather, its alteration into an object for display was an intentional, purposeful act. Layers of plaster or putty were packed into the skull's sinuses in order to seal off the bottom of the skull for its use as a vessel. As a result, it is unbalanced and bottom-heavy. The lacquer that coats the "bowl" was similarly applied in layers, as was the varnish coating the exterior. Transforming the skull in this way was not an "in the moment" decision but rather an act that required planning, time, and considerable work. These remains have been dehumanized in order to create a vessel that would visually symbolize or communicate something about the physician on whose desk it sat for decades. Some physical evidence remains of the skull's use as a desktop receptacle, including the angled cork placed in the foramen magnum and the few flecks of red glitter visible against the surface of the cranial cavity. This evidence is in keeping with the limited information we have available regarding the skull's history.[5]

The skull is a powerful and at times painful reminder of how medicine frames patients, the patient-physician relationship, changing attitudes regarding the treatment of human remains, and the care of marginalized patients. It also reflects the complicated history of these relationships and attitudes, as well as how these forces shaped the lives of individuals. Autopsies and anatomical dissection were and continue to be expected, acceptable activities for medical students. Cavalier treatment of human remains, as well as pranks, posed photographs, and the retentions of "souvenirs" from dissected cadavers, was also common practice.[6] While it is possible that the skull's person was part of a body donor program, they were most likely an indigent person who died in a public hospital or ended up in the city morgue. The body was probably unclaimed, or the person's loved ones were unable to pay the costs required to take custody of the remains. During this time, people may have been aware that having a loved one die in a public or charity hospital or leaving a relative's burial to the state could result in the body being used for dissection, but they typically did not have any other options.[7] Even if the skull's person volunteered their remains to further medical education (a rarity ca. 1920), surely they did not consent to have their skull transformed in this way, displayed in a physician's office, and finally transferred to a museum collection via a yard sale.

Given the limited provenance of the skull, the knowledge that it was likely acquired and transformed into a desk ornament without the consent of its person, what do we do with it now? As a curator, I have a professional and ethical responsibility to maintain the collections under my purview, make them appropriately accessible to the public, and generate and disseminate new knowledge based on the collections.[8] So while it is all well and good that I stay apprised of new developments in museum ethics, participate in various working groups focused on ethics and human remains, and write contemplative essays such as this one, I must also address the day-to-day treatment of the skull and plan for its future. Thus the decision to move it to its current storage location and continue to administratively address the skull as a collection item.

It is my personal and professional opinion that historical collections should be as accessible to the public as possible and, in general, if an artifact can be put to good use, it should be put to work rather than maintained in storage. Sometimes determining the best application of an artifact is straightforward. Other cases are much more complicated and potentially controversial to apply or interpret, thus many artifacts that could help provide insight into unsettling but important aspects of the medical past remain on a shelf. Deciding how to address the exhibit, research, and interpretive potential of an artifact is a complex and imperfect process.

We must think about the likely interpretations of the potential audience for the exhibit, online database, workshop, or class. We are not always equipped to provide the historical context necessary to avoid sensationalism of the item. Much to the chagrin of many a wordy curator, casual museum visitors typically do not read exhibit labels and other signage when visiting an exhibit. So one cannot rely on interpretive text to put artifacts into perspective if it requires a lengthy explanation.

Another consideration is whether the potential "shock value" or ethical transgression of exhibiting an item outweighs its educational impact. In the context of history of medicine museums, particularly one like the Dittrick that focuses on medical technology, it is my opinion that the educational goals of public exhibitions can be met without the display of human remains. In many cases, there are items within the collection that can more effectively convey a historical reality or concept to visitors. Thus, our current policy is not to include human remains in our exhibits.

While the Dittrick will not exhibit the skull during my tenure, it is still part of our collection and must be maintained as such. Like many of my colleagues, I question the ethics of museums amassing large collections only to keep the bulk of them in storage, out of the public eye and often inaccessible to all but carefully screened researchers. So even though there are artifacts that I will not display for ethical reasons, there are some that I will employ as an educational tool in certain contexts. As a former hospital nurse and current educator of medical and other clinical students, I thought perhaps the skull would meet an educational need that could help improve ethical and equitable delivery of health care in the future.

So far, small workshops with ten to twenty students and their instructors have been my most effective approach to improving understanding about the vestigial effects of the professional values of clinicians in the past. During these sessions, I walk (or prod) medical students through a visual analysis of collection artifacts while presenting a narrative about each artifact, usually the experiences of an imagined patient. Students are typically highly engaged in the discussion, and the questions raised by the object tend to spark important conversations between students about the historical roots of the health inequality they witness regularly.

Recently, I have employed the skull in such workshops as evidence of the often disrespectful, cavalier treatment of the dead in medical schools and the contempt some physicians had for their patients, particularly poor ones, in the not-so-distant past. After I share the known history of the skull, students observe the visible evidence of its intentional transformation into a decorative vessel and the physical signs of its use as a desk ornament. Next, I ask the students to examine the skull through a clinical lens, which encourages them to make a connection between the probable lived experience of the skull's person and their remains. Many students are visibly moved by witnessing the skull, while others seem to find it upsetting or unsettling. The student response to these workshops has been overwhelmingly positive so far, though I continue to evaluate their effectiveness and appropriateness. Attendees have approached me afterward to share how our discussion of the skull helped them understand the importance of medical history in their education or reconsider their relationship to the cadavers they dissect in anatomy lab. They frequently note that they had not previously considered that their instructors and clinical supervisors were taught by physicians from the same generation as the doctor who kept the skull on his desk for decades. I have no means of measuring the efficacy of my workshops beyond these volunteered responses, but I do feel that the discussion of the skull, in particular, will have an impact on how these new clinicians treat their patients.

My use of the skull in this workshop for clinical students will inevitably be questioned, which of course makes me worry about how readers will respond to this essay. Knowing the positive impact these workshops have had for the clinical students in attendance, I try to live with the discomfort that not everyone will agree with my choices. It is difficult to remain open to feedback when I feel so strongly about ethical approaches to history and museum work, and when I feel I must defend the decisions of my predecessors, who were working under a different set of institutional goals and values. The impulse to defend the integrity and authority of a museum is strong within the profession, especially in light of recent reckonings related to human remains. Often, rather than a careful, long, and open process of assessing difficult elements of a collection, the response is a knee-jerk reaction designed to address ethical problems (or hide them) before the institution gets "caught" with human remains or other troubling materials.[9] There are other instances where a museum's transparent, collaborative approach to collections stewardship has resulted in positive outcomes, such as repatriation of stolen remains or an exhibit cocurated by the museum and the descendants of Indigenous makers.[10]

In my opinion, ethical stewardship requires that museum professionals and their institutions allow themselves to be vulnerable to potential criticism, retaliation, and other consequences. This is a big ask given the economic vulnerability of many working in the field and the larger responsibilities of museums, libraries, and their parent institutions. We also make ourselves vulnerable to any upsetting feelings and a wide range of emotional responses, often intense, that can sometimes arise when working with human remains, especially those that seem desecrated. And while it has been my experience that parent institutions and professional colleagues are helpful and supportive under such circumstances, this is not necessarily the norm. It is important to remember that most museum professionals are doing the best they can with the resources available, and offer understanding and support, rather than assuming curators are untroubled by the human remains and other ethically fraught components of the collections they care for.

Given these very real and challenging limitations, moving toward more ethical stewardship practices is daunting. Fortunately, there are small steps underresourced repositories can take, including accepting our institutional and individual limitations. For example, there is no blanket policy that will adequately address the ethical concerns related to every item in the collection; this must be an ongoing activity. Best practices, ethics, and values, both personal and professional, will change over time, and our choices will be questioned. But what we choose to do now does not decide what we must do with these items in the future. Making difficult stewardship decisions can be paralyzing if we conceptualize our actions as setting a precedent that must be adhered to with all artifacts and for all time. This is not true. Not only will codes of ethics and best practices be different in the future, but educational needs, available resources, and institutional purpose will also have changed, often counter to our best predictions.

In my own work as a curator, I often worry about what my colleagues will think of my choices; whether I am doing right by a decedent, maker, or donor; and whether I am acting in the best interest of my institution, researchers, and the public. Often, public conversations regarding the preservation and display of politically or ethically divisive artifacts are simplified into a false binary of destruction versus celebration of the values represented by the artwork or artifact in question. This is unproductive, especially for those of us tasked with the actual preservation and display of the problematic artifacts. Any decision I must make regarding the troubling "candy dish" skull is not an all-or-nothing choice between destroying it and exhibiting it as the centerpiece of a macabre-themed anatomical display. There are many options. When working through ethical dilemmas, I remind myself that I have the right to change my mind based on new evidence, experience, or knowledge. I must accept that I will make mistakes, that my choices will be questioned, and that there is no perfect solution. This is not very comforting, and I am fortunate that I do not make these decisions alone. There are many experts from a range of fields willing to help their colleagues address problematic collections.

The skull will become increasingly fragile in the coming years. This is due to the amateur preparation of the skull, which did not properly remove soft tissue and rendered it structurally unsound. In addition, the use of varnish and lacquer on the surface has sealed in the fatty, oily components of living bone that, with proper preparation and storage, would have gradually dried out over time. There is little that could be done in terms of conservation treatments to prolong the stability of the skull. Thus, there will come a time when the skull falls apart, regardless of how carefully we handle and store it. In pieces, it will no longer be visually understandable as a desk ornament or ashtray and will no longer "work" as physical evidence of the often contemptuous attitudes toward the indigent dead held by many early twentieth-century clinicians. At that point, my current plan—which I reserve the right to change—is to work with the university anatomy lab chaplain to have the skull remnants cremated as part of the annual memorial service held in honor of those who donated their remains for medical research. This is not a perfect solution, but it feels like the best thing we can do right now.

NOTES

1. On the history of estimating the sex and reconstructing the gender of archaeological human remains, see Molly K. Zuckerman and John Crandall, "Reconsidering Sex and Gender in Relation to Health and Disease in Bioarchaeology," *Journal of Anthropological Archaeology* 54 (2019): 161–171.

2. Marie C. Malaro and Ildiko Pogány DeAngelis, *A Legal Primer on Managing Museum Collections*, 3rd ed. (Washington, DC: Smithsonian Books, 2012), 83n70.

3. Leonie Hannan and Sarah Longair, "Developing a Methodology: Understanding Museum Collections and Other Repositories," in *History through Material Culture* (Manchester: Manchester University Press, 2020), 70–94. For insight into the complexity of provenance research, see Shelley Saggar's essay in this volume.

4. I am grateful to Nicole M. Burt, curator of Human Health and Evolutionary Medicine at the Cleveland Natural History Museum, for her unofficial yet insightful forensic examination of the skull as well as her advice regarding the ethics of human remains in museums.

5. At the time of its donation, both the donor and museum staff envisioned the future of the skull as a powerful teaching tool. At no point was there an intention to publicly exhibit the skull.

6. John Harley Warner and James M. Edmonson, *Dissection: Photographs of a Rite of Passage in American Medicine, 1880–1930* (New York: Blast Books, 2009).

7. Michael Sappol, *A Traffic of Dead Bodies: Anatomy and Embodied Social Identity in Nineteenth-Century America* (Princeton, NJ: Princeton University Press, 2001).

8. American Alliance of Museums, *A Code of Ethics for Curators* (Washington, DC: AAM, 2009), https://www.aam-us.org/wp-content/uploads/2018/01/curcomethics.pdf.

9. See Jamie Jelinski, "Go and Take a Look at Millie Now," in *Museums and the Working Class*, ed. Adele Chynoweth (New York: Routledge, 2021), 74–87.

10. See Meghan Bill, "Tracing Threads of History: Rediscovering Indonesian Textiles at the Brooklyn Museum," *Museum Anthropology* 44, no. 1–2 (2021): 38–54.

CHAPTER TEN

Disability, Archives, and Museums

KATRINA JIRIK

Douglas Baynton, in his history of disability and immigration, observes, "For a long time, we [historians] never thought of disabled people, but they are everywhere, and our histories are defective without them."[1] And being defective makes our histories, and the archives and museums where historians work, full of ethical issues, because as Manon S. Parry, a historian of medicine and exhibition curator, and coauthors state, "the cultural spaces of museums and archives are a primary arena in which society's values are defined, preserved, reified, and shared, shaping who is valued and included in a country's notion of citizenship."[2] For too long, history was the story of powerful white men who left behind extensive written records. The introduction of histories of marginalized populations, such as women's history and African American history, led to an intersectional and nuanced understanding of the past.[3] And yet, very little research has been done on the largest marginalized population, disabled people, who, in the United States, make up 25 percent of the population.[4] This silence is a manifestation of the marginalization of disabled people and represents major ethical challenges in our work as historians, especially historians of medicine.[5] In her contribution to this volume, Ayah Nuriddin makes the point that historians need to wrestle with the difficult questions of who is and is not included, and how that shapes our work. While the disability population is diverse, like all marginalized populations, and does not speak with one voice, collaborating with disabled people is imperative in beginning to address the ethical challenges of marginalization.

It is not simply the history of disabled people that is sorely lacking; the use of disability as an interrogatory tool to help investigate the intersectionality of historical events is also incompletely theorized.[6] Concepts of fit and unfit bodies, used as ways to justify discrimination, have a solid place in American history. Enslaved African Americans were counted as three-fifths of a person in the US Constitution. They were portrayed as having characteristics that exemplified low intelligence, laziness, and bodies that were not normal, all of which were used to justify slavery. Women have been denied equal citizenship rights based on their purported lower cognitive skills and fragile nature, which made them suited only for tasks within the domestic sphere. Women attempting to break out of these stereotypes were often deemed irrational. All these characteristics, however, are characteristics associated with disability, with unfit bodies.[7] How does using a disability lens to analyze "social relationships, legal institutions, democracy, education, medicine, bodies, epistemological frameworks, foreign policy, social welfare, and on and on" change our perceptions?[8] The use of disability as an analytic tool must not supersede the establishment of disability history, or there is a risk of losing the lived history of disabled people. This is where the ethical work of historians, as researchers and teachers, becomes important.

As historians, we gravitate to two locations, archives and museums, which in many ways are our natural homes, places to do the hard work of producing knowledge. Right now, ableism is embedded in spatial accessibility. Problems with physical access to archives and museums have been detailed in many publications. As a wheelchair user, however, I think it is an ethical issue when the way I get into many buildings is by the trash cans, something Nicole Schroeder also mentions as a common occurrence in her contribution to this volume. The response (excuse) that at least I can get into the building does not begin to address the lack of value given to my participation in those locations. I am an academic researcher. Locating the entrance I must use by the trash cans conveys the message that my value is equivalent to that of the trash. More importantly, however, is the ableism embedded in knowledge production. Too often the voices of disabled people are missing from the archival record, muted, silenced by the voices of prominent actors or legal prohibitions, or hidden or misclassified in finding aids. And yet the recovery of this information is vital to our understanding of how past beliefs and actions influence our current laws and policies. As historians, we need to dig deeper, past the preserved documents of prominent actors, to those whose voices are harder to find but offer different ways of understanding the past.

Archives

Looking for disability in the archives is often a difficult task. The concept of disability has changed over time, from a moral issue to a medical issue to a social and political issue.[9] Many of the existing records are, in fact, medical records that document medical interventions and often histories of abuse of and violence against disabled people. They are filed by doctors, nurses, and aides and report their views, not the views of disabled people. While these records are part of disability history, they do not convey the lived experience of those with disabilities. They prompt serious ethical questions regarding whose voices are deemed important, and too often it is not the voices of those with lived experience. The reliance on medical and institutional records presents other issues. For example, where are the stories of disabled people who lived their lives in their communities, not in institutional or medical settings? Where is the information on the adaptations they made to successfully live their lives? Institutions for the "feebleminded" in the United States held, at most, 5 percent of the total number of people considered "feebleminded." How do we access the stories of the other 95 percent? Our histories are vastly skewed when we rely only on the medical and institutional records that are more readily available, and such histories are ethically fraught, as they make invisible the stories of a vast number of people.

Archivists play a critical role in interpreting source material because they determine what gets preserved, what gets tossed out, and how material gets classified. With those determinations, archives become sites of power and control and questionable ethics.[10] As women's history became established, archivists had to go back and reindex collections to consider women's history. The same thing needs to happen with disability history. It needs to go much further, however, because so much information is not currently contained in archives. Oral histories need to be collected.[11] Outreach to disabled communities needs to be done to collect their stories and artifacts. Interviews need to be conducted to capture the human experience of disabled people and the people who care about them, something highlighted in the essay by Britt Dahlberg and Jessica Martucci in this volume. Outreach to small community archives may bring information about communities' interactions with disability. An online sourcebook of archives that contain disability information would also be helpful, as disability information does not always follow provenance rules and may be scattered in a number of places. Without these efforts, archives will continue to be places that deny disabled people their place in our common history.

An additional archival issue is access to records that may be sealed. States have different parameters regarding privacy. Especially with the Health Insurance Portability and Accountability Act of 1996, established to protect sensitive patient health information from being disclosed without the patient's consent or knowledge, medical records tend to be sealed for a period of time, often seventy-five years, or in some cases permanently, unless a court order is obtained.[12] While this protects privacy, it also makes research difficult. Given the history of abuse of disabled people, sealed records can be instrumental in hiding that abuse.

Additional harm may happen to disabled researchers when the records are heavily tilted to accounts of abuse, violence, or total erasure from the historical record. According to Gracen M. Brilmyer, archival "data illustrates the prevalence of disability stereotypes, tropes, and limited perspectives within the records that document disabled people. . . . Witnessing the violence of the past is emotionally difficult for many disabled people."[13] This is reflected in the often expressed thought, "If I were alive at that time, it could have been me," or, for more current events, "There but for the grace of God go I." We also need to see archival sources that tell the stories of successful disabled people, because in many cases their success was due to their disability, not in spite of it. For example, Wanda Diaz-Merced is an astronomer. She developed a new way to study space called sonification, which has helped her make new discoveries in light from visible stars, information that computer processing can sometimes obscure. She is also blind.[14]

Are there collections within archives that are undocumented because archivists viewed disability as only a medical issue or social barrier and not as an identity or experience?[15] Who covers the costs of integrating disability into the archives, especially since cost is often cited as the reason disability cannot be addressed in public spaces? Where are the archivists with disabilities who have both archival and lived experience that will allow them to critically examine archival material through a dual lens? How can we, as archivists, historians, and teachers, use a disability focus to reexamine the history we write about? What are the ethical costs if we choose not to address these issues?

Museums

Another place familiar to historians is museums. Some have undergone a change in focus in recent decades. They have moved away from the collection of objects and a neocolonial legacy to a focus on the complex relationships between objects and people. Yet museums tend to represent the dominant cultural values of society, subordinating or rejecting the values of marginalized populations.[16] The International Committee for Museum Management states, "It is a fundamental responsibility of museums, wherever possible, to be active in promoting diversity and human rights, respect and equality for people of all origins, belief, and background."[17] This is a call for an ethical reconsideration of museums' focus, as museums need to be agents for change, or they risk perpetuating and affirming the undervalued social positions of marginalized groups. Some museums are, increasingly, offering exhibitions and activities that target diverse audiences. There is a growing push to acknowledge marginalized populations in the work museums do.[18] Yet with all the talk of museums being more participatory institutions, control still remains with the institutions. While much is made of museums' outreach to marginalized populations, much of the outreach lives in the publicity department, not in actual programming.[19] While progress has occurred for some marginalized racial and ethnic populations, disabled people have been largely ignored. Shelley Angelie Saggar explores an aspect of this in her essay on the decolonization of museums in this volume.

Disability, by its very name, is placed in a position of inferiority in relation to the standard of normality. In the media, various stereotypes predominate: pathetic, violent, evil, a burden, super cripple, and incapable of living in society, among others.[20] These cultural values, as presented by museums, contribute to the discrimination faced by disabled people in society and are an ethical challenge. Too often it is museum staff who speak out on issues related to disability, which perpetuates the disenfranchisement of disabled people as it is not their voices that get heard but the voices of people who think they know what disabled people think and want. The agency of museum staff is important, but it should not be prioritized at the expense of those with lived experience. Frequently, museums respond to the needs of marginalized populations by enlisting them in a new category, "the vulnerable," adding another form of stigma and alienation from the dominant society. Disability is intersectional, and the discrimination faced by marginalized people is amplified when disability is added to a person's identity. As the primary audience of museums is educated and professional people, until recently exhibits have catered to the predominant values of these groups, rejecting the values and lives of marginalized communities, especially disabled people.[21] Rosemarie Garland-Thomson, a disability historian, focuses "on how public cultural narratives on disability limit the way disabled people are seen."[22] Uncomfortable with how to proceed, curators worry that they will offend either people with disabilities by poor language choices or the able-bodied audience whom they perceive as having little interest in disability-related exhibits. This often leads to stereotypical exhibits on how awful having a disability is or the super-crip who has overcome the challenges of living with a disability. These efforts create a misrepresentation of what life is like with a disability and limit societal opportunities and social status for disabled people.[23] Instead of doing the ethical work to make museums more inclusive, many do nothing, continuing to perpetuate the silencing of disabled people. The claim that a museum failed to address the ethical challenge because of worries about getting it wrong is a cop-out because, as noted by Patricia Roque Martins, integrating disability into museum exhibits seems to engage visitors in understanding the historical and cultural effects that have shaped a negative perception of people with disabilities.[24]

As most employees in museums do not claim disability status, currently the only way to counteract these negative perceptions is for museum staff to commit to a fully collaborative process with disabled people to reform the museum. This collaboration needs to involve much more than allowing disabled people into the curating space of the museum. There needs to be collaborative power sharing, something most museum staff find challenging because many disabled people do not have the accepted credentials for curation. In curating artifacts related to disability, it is critical to have someone with lived experience participate in the curation to make sure it is accurate. For example, the Pennhurst Museum, which is located on the grounds of

A cabinet of common, colorful children's toys in reds, yellows, oranges, and greens, which contrast starkly with the drab gray of the interior of the Pennhurst Museum that formerly housed the Pennhurst State School and Hospital.
Photo by Nathan R. Stenberg. Original in color, reproduced in black and white.

the former Pennhurst State School and Hospital in Pennsylvania, displays artifacts collected from the former institution for cognitively impaired people. One of the displays is a grouping of toys.

Curators without disability experience might erroneously assume they were items of comfort to the people who lived there. Nothing could be further from the truth. In *Public Hostage, Public Ransom: Ending Institutional America,* by William Bronston, there is not one picture of a resident with a toy.[25] Toys did not belong to the residents; they were objects of coercion belonging to professionals and were used as a temporary reward for completing the task demanded by the professional. The reward was often fleeting, as the toy was reclaimed by its owner and kept until another task was completed. The resident never got to keep the toy. This is a critical aspect in elucidating institutionalization and yet easily missed by curators without lived experience.

An example of the incorporation of curation by people with disabilities into museum exhibits is the Rethinking Disability Representation in Museums and Galleries project in nine museums in the United Kingdom, an endeavor of the Research Centre for Museums and Galleries (RCMG). This project included "disabled activists, artists and cultural practitioners" who helped guide the museums as they developed exhibits and programs representing disabled people's lives.[26] Although the museum projects varied (e.g., exhibitions, workshops, displays, and films), the overall response from visitors was positive, indicating that disability was not only a topic of general interest but also one they could engage with. What visitors particularly liked was the use of real disabled people's voices and lived experiences. For example, when a call went out for people to bring adaptive items to Colchester and Ipswich Museum for a display, a woman brought a tube of colorful lipstick. She wore the lipstick because it helped her visually impaired daughter recognize facial expressions.[27] One of the results of this project was increased recognition of the social barriers and stigma faced by disabled people, and it allowed visitors to see disabled people in new ways, specifically by perceiving that disabled people were "an integral part of society and should be displayed as such rather than hidden away."[28] Another result was to highlight issues disabled people face that are barriers to their participation in society. Visitors became more aware of the marginalizing effects of service provision inadequacies and funding issues.

It is not only museum visitors who have experienced changes in attitudes; museum professionals have too. Heather Smith of the National Trust states, "We cannot continue to exclude disabled people from the heritage we have. I can see so many items in our properties connected to disability history and we just do not tell anyone about them yet."[29] Claire Madge, founder of Autism in Museums, believes that "recent work at the Wellcome Collection . . . has shown how museums and galleries can work with groups and individuals traditionally seen as 'outsiders' to make sure the whole spectrum of voices and experiences are heard."[30] Even curators have experienced changes in their perspectives after participating in RCMG projects. Lauren Ryall Waite comments, "Through working with RCMG at Thackery Medical Museum I became more aware of the lack of representation in our collections of disabled voices and of the bias of the medical model of disability. Work with RCMG has led to a personal change in how I approach medical collections as a curator."[31] The responses to these projects suggest they may result in transformations necessary to counteract the ethical challenges of negative social perceptions of disability.[32]

Medical museums present a huge number of issues when considering disabilities. The artifacts are generally labeled as specimens when, in reality, they are body parts, often human body parts. The term *specimen* disengages the viewer from the person the body part belonged to. The humanity of a person is compressed into the specimen on display. Other aspects of the person's life are made invisible, reinforcing the medical model that says the most important thing about disability is that it must be fixed or cured. It negates, often through the curation, the life lived by the person, how the person accommodated the disability, and how they interacted with and were interacted with by society. From personal experience, I can attest that there are even more negative outcomes for disabled people. I was in the lobby of the Mütter Museum when a tour group of older elementary students (fifth or sixth graders) was leaving the museum.[33] From their actions and stares, I realized I was being perceived as a specimen, not as a human being using a wheelchair. Because the museum had disengaged the exhibited body parts from the human experience, the students had difficulty integrating the divide presented by the Museum between specimens and real disabled people, when coming into contact with one. It facilitated the idea that a disabled person is an "other," someone outside of their experience, enabling discrimination. The Mütter Museum had just taught children that disability is an abnormal condition that places a person in the category of "an other." It taught them that discrimination was acceptable. There are serious ethical challenges that need to be addressed here.

It does not need to be this way. Medical museums like the Mütter Museum could use curation to discuss a person's experience.[34] They could hire people with disabilities to be tour guides to help visitors, especially children, learn that disability is a part of life. They could stop fundraising from Halloween nights that portray the body parts, and therefore the people associated with them, as objects of horror. They could actively recruit and pay disabled people to curate the artifacts from a disability perspective. They could make sure visitors were taught that disability is part of the human condition, and that civil and human rights apply to everyone. Without these changes, medical museums remain an ethical challenge.

Conclusion

If we truly wish to do less harm, then it is incumbent on us to work to make disability an integral part of our archives and museums. They are where we, as historians, do our work. Disabled people are the largest marginalized population in the United States and yet they are the least represented. The inclusive effort needs to move far beyond physical and sensory access. It needs to bring the lived experiences and the voices of disabled people to the forefront. Disability needs to be not only a respected topic of research but also an interrogatory tool to examine our varied histories, not just our medical histories. Without critical changes, our histories will be defective, submersed in an ethical miasma that taints our work.

NOTES

1. Douglas C. Baynton, *Defectives in the Land: Disability and Immigration in the Age of Eugenics* (Chicago: University of Chicago Press, 2016), 138.

2. Manon S. Parry, Corey Tijsseling, and Paul van Trigt, "Slow, Uncomfortable, and Badly Paid: DisPLACE and the Benefits of Disability History," in *Museums and Social Change: Challenging the Unhelpful Museum*, ed. Adele Chynoweth et al. (London: Routledge, 2020), 149.

3. Authors such as Susan Burch, in *Committed: Remembering Native Kinship in and beyond Institutions* (Chapel Hill: University of North Carolina Press, 2021), and Adria L. Imada, in *An Archive of Skin, an Archive of Kin: Disability and Life-Making during Medical Incarceration* (Berkeley: University of California Press, 2022), address marginalized populations.

4. There is an ongoing discussion within the disability community about whether person-first language (*person with a disability*) or identity-first language (*disabled person*) is preferable. There tends to be a generational divide, with older people using person-first language and younger people using identity-first language, but this is not always the case, and it is considered best practice to ask a person what they prefer.

5. See Ayah Nuriddin's essay in this volume for a larger discussion of silences and race. Many of the topics relate to disability as well.

6. Kim E. Nielsen, "Historical Thinking and Disability History," *Disability Studies Quarterly* 28 no. 3 (Summer 2008), https://dsq-sds.org/index.php/dsq/article/view/107/107.

7. Nielsen; Baynton, *Defectives in the Land*, 8–17.

8. Nielsen, "Historical Thinking."

9. Bridget Malley, "Documenting Disability History in Western Pennsylvania," *American Archivist* 84, no. 1 (Spring/Summer 2021): 19–20.

10. Joyce Gabiola et al., "'It's a Trap': Complicating Representation in Community-Based Archives," *American Archivist* 85, no. 1 (Spring/Summer 2022): 60–62.

11. Oral histories of scientists with disabilities are available in "Science and Disability Oral History Project," Science History Institute Museum and Library, accessed November 15, 2024, https://digital.sciencehistory.org/collections/t28xqhe. Survivor oral histories are used in Darby Penney and Peter Stastny, *The Lives They Left Behind* (New York: Bellevue Literary Press, 2009), Correspondence Files, 1982-2014 B2838-21 3 cubic feet Willard State Hospital Patient Case Files, 1900–1990, bulk 1940–1980, https://www.archives.nysed.gov/research/new-accessions-july-2021. Imada, *Archive of Skin*, also makes use of oral histories.

12. "Health Insurance Portability and Accountability Act of 1996 (HIPAA)," Centers for Disease Control and Prevention, September 10, 2024, https://www.cdc.gov/phlp/php/resources/health-insurance -portability-and-accountability-act-of-1996-hipaa.html.

13. Gracen Mikus Brilmyer, "'I'm Also Prepared to Not Find Me. It's Great When I Do, but It Doesn't Hurt If I Don't': Crip Time and Anticipatory Erasure for Disabled Archival Users," *Archival Science* 22, no. 2 (2021): 167–188, quotation on 168.

14. "Wanda Díaz-Merced," Royal Society, accessed October 26, 2022, https://royalsociety.org/topics-policy/diversity-in-science/scientists-with -disabilities/wanda-diaz-merced/.

15. Sara White, "Crippling the Archives: Negotiating Notions of Disability in Appraisal and Arrangement and Description," *American Archivist* 75, no. 1 (Spring/Summer 2022): 123.

16. Jamie Allen Brown, "The Case for the Inclusive Museum: A Perspective from Excluded Groups and Communities," in *Papers from the ICOFOM Symposia in Buenos Aires, Rio de Janeiro and St. Andrews*, ed. Bruno Brulon Soares, Karen Brown, and Olga Nazor (Paris: ICOM/ICOFOM, 2017), 120.

17. Bernadette Lynch, "Introduction: Neither Helpful Nor Unhelpful—A Clear Way Forward for the Useful Museum," in *Museums and Social Change: Challenging the Unhelpful Museum*, ed. Adele Chynoweth et al. (New York: Routledge, 2021), 11.

18. Lynch, 8.

19. Lynch, 10.

20. Patricia Roque Martins, "The Representation of Disability in DGPC Museums Collections: Discourse, Identities and Sense of Belonging," in *Representing Disability in Museums: Imaginary and Identities*, ed. Patricia Roque Martins, Alice Lucas Semedo, and Clara Frayao Camacho (Porto, Portugal: CITCEM, 2018), 15.

21. Brown, "Case for the Inclusive Museum," 120.

22. Martins, "Representation of Disability," 16.

23. Parry, Tijsseling, and van Trigt, "Slow, Uncomfortable, and Badly Paid."

24. Martins, "Representation of Disability," 8.

25. William Bronston, *Public Hostage, Public Ransom: Ending Institutional America* (Conneaut Lake, PA: Page, 2021).

26. Richard Sandell and Jocelyn Dodd, "Activist Practice," in *Representing Disability: Activism and Agency in the Museum*, ed. Richard Sandell, Jocelyn Dodd, and Rosemarie Garland-Thomson (New York: Routledge, 2010), 13.

27. "Life beyond the Label: Rethinking Disability Representation," Colchester and Ipswich Museum Service, accessed November 12, 2022, https://le.ac.uk/-/media/uol/docs/research-centres/rcmg/publications /colchester-and-ipswich-museums-service.pdf.

28. Jocelyn Dodd et al., *Rethinking Disability Representation in Museums and Galleries: Supporting Papers* (Leicester: RCMG, University of Leicester, 2013), 47.

29. Emma Shepley, *Advancing Disability Equality through Cultural Institutions: Research Impact Report* (Leicester: RCMG, University of Leicester, 2013), 16.

30. Shepley, 28.

31. Shepley, 32–33.

32. There is an effort underway to establish a national museum of disability history and culture, which would be associated with the Smithsonian. Some of the people involved in the effort are members of the Pennhurst Memorial Preservation Alliance, and more information can be found on their website. The Smithsonian is also involved in this effort.

33. The museum has pulled all its videos while it reassesses ethical issues related to the display of human body parts. This has generated a large public backlash. Michael Tanenbaum, "Mütter Museum Weighs Ethical Concerns over Online Exhibits Displaying Human Remains," Philly Voice, May 12, 2023, https://www.phillyvoice.com/m%C3%BCtter -museum-philly-online-exhibits-human-remains-ethics/.

34. See Melissa Grafe's essay in this volume for information on the Mütter Museum. See also Aparna Nair, "The Mütter and More: Why We Need to Be Critical of Medical Museums as Spaces for Disability Histories," Disability Visibility Project, July 28, 2023, https://disabilityvisibility project.com/2023/07/28/medical-museums-disability-power-and-empire/; Maura Judkis, "A Museum's Historic Human Remains Are Now the Center of an Ethics Clash," *Washington Post*, July 26, 2023, https://www .washingtonpost.com/lifestyle/2023/07/26/mutter-museum-controversy -philadelphia/; and Ken Knickerbocker, "Washington Post: Historic Human Remains Create Ethics Clash at Mütter Museum," MONTCO Today, August 1, 2023, https://montco.today/2023/08/historic-human -remains-mutter-museum/. The museum has begun to include some

narration about the lives of the disabled people whom its exhibits document. But in a YouTube video, Mütter educator Marcy Engleman states that she hopes visitors have a "disturbingly informative experience." Localish, "Mütter Museum Has Einstein's Brain and Other Creepy Exhibits | Secretly Awesome," YouTube video, 3:29, December 1, 2018, https://www.youtube.com/watch?v=32UEv6B7E8o. Also on YouTube is a video that describes the Mutter as a place for people "searching for the macabre" that is full of "eccentric curiosities of the medical profession" and "downright bizarre." These comments are all ableist. Travel Channel, "Philadelphia Mütter Museum | Travel Channel," YouTube video, 2:25, April 18, 2014, https://www.youtube.com/watch?v=9GZIfOXKE7o.

CHAPTER ELEVEN

The Commercialization of Remains and Records

APARNA NAIR

Like many historians, I had long been an occasional collector of ephemera and objects that fascinated me—postcards, posters, comics, advertisements, old books on various topics related to my research or teaching. In 2017, this desultory collecting practice became more serious and organized when I began creating mini exhibits with students on a range of topics and turned to local secondhand and junk shops as well as digital storefronts like eBay and antique stores. These exhibits and the collecting-based research I did, and continue to do, have been truly effective in ways beyond my own expectations; students find the use of material culture revelatory in accessing the lived experiences of historical subjects but also enjoy the thrill of the "hunt" for ephemera or other objects for our exhibits. This method has also been generative for my own research, and I currently have five ongoing mini collections on the following topics: disability histories in British India, vaccination documentation and ephemera, guide dog histories, seals and stamps, and comics and public health, all of which have fed into different parts of my research and writing.

Collecting as practice has been the subject of extensive research attention across the decades. Walter Benjamin famously described "collecting [as] a form of practical memory" and also invited readers to reflect on collecting as he unboxed his own personal library.[1] In the decades since Benjamin, many scholars, including historians, have explored why we collect—whether for pleasure, science, or "obsession"; as performance of economic or social status; as institutional practice in archives and museums; as an act of "self-conscious preservation of a lost or disappearing world"; as an act of rescue; or as a way for marginalized populations to change the narratives told about their experiences and histories.[2] Even a cursory glance at the wider histories of medicine demonstrate how important personal collecting practices have historically been in shaping the library or archive, whether we consider the early modern cabinet of curiosity or nineteenth-century physicians' tendencies to accumulate personal collections of pathological specimens.[3]

So what ethical concerns could arise when historians of medicine, public health, disability, or the body more broadly turn to collecting as intentional historical *method* instead of as desultory practice, hobby, or consumption? Certainly, there are arguments that could be made using feminist, queer, and crip theories for researchers to seek out sources through collecting.[4] Collecting as method has served to counter state priorities and oppressions or socially enforced silences: notable examples of community-created archives of queer life and history include the International Gay and Lesbian Archives and the Lesbian Herstory Archives in the United States.[5] Franklin Robinson Jr., has described those organically produced personal collections of papers, objects, memorabilia, and photographs that accumulate out of people's interests, experiences, and networks (in this case, constituting the archival record of LGBTQ+ generations) as the "archival closet."[6] Robinson also tracks how such personal collections become institutional: for instance, a collection of photographs of an unidentified gay World War II veteran that were purchased at a yard sale and then broken up by an eBay seller and sold as separate images, which were then bought up by the National Museum of American History. For disabled scholars, collecting could offer one way to counter the exclusionary and still largely unaddressed ableism that restricts their access to traditional archives and libraries.[7] Certainly, for me, the constraints of my own epileptic bodymind and the inaccessibility of so many archives and libraries have been a part of my reason for turning to collecting and digital archives in the past seven years.

In this essay, I interrogate the collecting practices I have engaged in while teaching and dealing with objects, corporeal traces, and body parts in commercialized settings like eBay and other spaces. I approach these ethical issues both at the meta level (How do we reckon with the broader issues of provenance and the questions of the harms we might perpetuate by participating in the commercialization of texts, objects, and ephemera?) and at the level of the individual (How do we strive for ethical conduct in our everyday practices of collection and curation, and what do we learn when we find ourselves confronted by unethically procured objects?). In the first case study, I explore how historians might confront unethical practices regarding the sale of human body parts. In the second case study, I discuss what happened when I found the medical records of disabled children for sale on eBay and argue for an ethics that interrogates and challenges the weaknesses of international, national, and state laws and the failures of institutions. I suggest that there are both general and unique ethical considerations that historians of medicine need to examine when we are "collecting, contributing to, and creating the archive ourselves" using eBay or other platforms.[8] I draw partly on a method from historical geography that Dydia DeLyser has described as "archival autoethnography," but I also write this as both a meditation and a reminder to myself about the possibilities, ethical concerns, and risks of collecting as method.[9]

Collecting offers a range of possibilities for historians of medicine, no matter what your interests. Taking eBay alone, textual sources are incredibly diverse and range from indenture documents to tuberculosis seals, to antivaccination ephemera, to newspaper articles marking events in the history of medicine, to personal paper collections, and to old books. You could find anything from tiny glass-eye wash cups to nineteenth-century amputation knives, to antique bath chairs, to antique quack medicine bottles, to twentieth-century "electrical medicine" kits, to homemade prosthetics. What, then, are the procedural and ethical concerns that arise from the practice of collecting such objects? Using the platform for research can be complicated. eBay is often popularly described as an "archive," and it certainly has some elements of a more traditional archive, but its organizational systems and categorizing practices are vastly different, more untrustworthy, far more chaotic, and shaped by the historical narratives and cultural meanings ascribed to objects and documents by sellers. Describing her search for *cartes de visite* on eBay, Zoe Trodd notes that eBay reminds her of the *Wunderkammern*, a "world of wonders in one cabinet shut."[10] But unlike the traditional brick-and-mortar library or even the digital archive, eBay offers "randomness and disorder" to the historian seeking finds, and as Cecily Devereux argues, this very randomness allows us to see how the wider public "orders" and categorizes objects, forms connections, and presents these objects and documents as part of a historical narrative.[11]

But this is not the only ethical challenge for collecting-as-method. When we consider eBay in particular, it is clear that the platform is ephemeral and digital yet rooted in the material and physical storefronts; but these storefronts are scattered all over the world. Although registered in the United States, the platform allows buyers access to sellers situated across the globe. This has been useful for me, as a historian of disability and medicine in South Asia, and I have been able to find everything from a 1950s-era tuberculosis vaccination document for a child from Penang, Malaysia, to seals used to raise funds for disabled World War I veterans in British India, to an album of music by a blind Indian composer from the 1940s. The site has unquestionably globalized the possibilities of collecting, but with these expanded spatial horizons, the ethical issues that rise are complex. And as with the archive, and the museum, it is useful to think of eBay in terms of access and power. For the historian of medicine, eBay could work as a complementary (commercial) archive, especially important for independent or precarious scholars or those without institutional supports. But even so, the very act of collecting requires a certain degree of privilege—both financially and spatially. Better-funded scholars living in the Global North unquestionably have a degree of access to the global marketplace of commodified history that other scholars will never have.

Another issue is that of provenance. On eBay, there are few rules about establishing the histories and origins of objects and documents, which are generally erratically documented and sometimes missing, and are shaped also by seller reputation and histories. Despite this, I try to investigate what the seller includes in the description, and if I am familiar with the seller or am a return buyer, I sometimes ask where the seller acquired the objects or documents in question. But the majority of the materials in my collection have very sparse information on provenance, leaving me with many absences and omissions to navigate. For instance, regarding a handmade wooden leg that was crafted by a disabled Oklahoman sharecropper in the early twentieth century, I was able to get a detailed family history from his great-grandson, who found the leg in his parents' garage, along with additional information and materials, including photographs, although they were not able to tell me his name. At the same time, I have a wooden prosthetic hand from Morocco, which looks like it is from the mid-twentieth century, but I was unable to get any information from the seller, who had found it discarded and "rescued" it. Working with objects like this requires critical engagement with the chiaroscuro patterns of information on provenance, as well as attempts to make sense of these objects and their histories through additional research based in more traditional archives and libraries. As Gracen Brilmyer reminds us, archival materials are "always already incomplete," because they are the accretions of objects from different sources, often with "unknown authors, histories, and provenance." Brilmyer is referring to the traditional archive, but this insight could be used as a foundation to understand and work with collections too.[12]

One of the most obvious ethical issues with collecting using eBay is that the platform is slow to identify problematic sales and sellers. Ken Hillis has pointed out that eBay's current practices and ideals are unquestionably shaped by its libertarian founder, Pierre Omidyar.[13] Despite some efforts by the company to clean up its act, we continue to be confronted with objects or documents that we *know* have been procured through problematic routes. For example, I have routinely found sellers offering up documents from India that I know, from decades of experience in brick-and-mortar archives of the region, were taken, stolen, or lost. I have found official texts with state seals, which clearly belong in particular sections of regional archives. And it is not only these kinds of documents; there have been several documented cases of theft of materials from various libraries and archives, materials that were then sold off through eBay. This, then, is one of the most obvious ethical issues with collecting in general: such thefts are in part driven by the existence of a market for unethically acquired materials; and when I collect as a historian, I contribute to the perpetuation of this market.

In 2021, I came across one of the most egregious examples of such unethically acquired objects. Several sellers, mostly based in New Jersey and the surrounding area, offered hundreds of pages of individual medical records of institutionalized psychiatric patients from the 1940s onward, which included intensely personal details, such as their family and immigration histories, addresses, sexual histories, diagnoses, and treatments. Other records included ward journals from New York state psychiatric hospitals, "miscellaneous patient files," and admission cards with photographs. Some of the objects on sale were truly poignant, such as trinket boxes belonging to patients, but others, such as the signs from state psychiatric hospitals and morgue doors, straitjacket buckles, and master keys, symbolize the violence, suffering, and control embodied by the asylum. I was horrified, because I knew that there were legal protections that prevented the public from accessing these records, and reached out to a group of historians of medicine, including Kylie Smith, whose help was invaluable in the subsequent yearlong effort to get these records removed from eBay, which involved reporting these sellers to the state archives and following up with eBay itself. When Smith and I investigated these records, we realized that they had to have been taken from various institutions for disabled people that had been shut down in the late twentieth century in the United States, that these records should have been handed over to state archivists, and that there were multiple ethical and (potential) legal violations in question.[14] But what horrified us the most was the deeply troubling attitude to a dark history, perhaps best embodied by how one of these sellers advertised the abuse of disabled patients as a selling point, with a description that said that the patient had been "Slapped by Nurse."

But it is not only these kinds of documents and objects that we encounter while collecting histories of medicine. In the years after eBay's founding, people tried to sell several disturbing "commodities," including human remains, guns, and ammunition, on the platform before these categories were eventually banned.[15] But sellers continued to slip by safeguards, especially in the case of human remains. Newspaper articles show how sellers tried to sell a stolen Indigenous Hawaiian skull taken from an archaeological site in Kaanapali Beach, "a real human skull from the Korean war," "Inca skulls," and the skeletal remains of a mummified Scottish child from a nineteenth-century physician's collection.[16] To get a sense of the scale of this trade at the time, consider this case: in 2007 alone, the US National Park Service alerted the Louisiana State Division of Archaeology to the sale of what was suspected to be a human skull, offered by a Louisiana seller.[17] When the Louisiana Department of Justice then tracked the sale of human skulls on eBay for seven months, they found, in that short period alone, that 237 people had listed 454 skulls for sale, with bids ranging from $0.01 to upwards of $5,000.[18]

Today, skulls and other human remains might be slightly harder to procure through eBay, but it is still possible. Even as I write this piece, in August 2023, I was able to find a skull for sale by a UK-based seller, in the "Occult and Witchcraft" category. The seller had named the skull Nancy, because "her real name is very long and unpronounceable," which might suggest that the skull belonged to someone who was not British. The skull, which the seller claimed to have had in her possession since 1971–1972, also had clearly been prepared as a medical aid, as it had "springs on her jaws and catches on her skull." Most distressing of all, perhaps, is the fact that this skull was "bedazzled" with colorful and sparkly plastic gems scattered all over the eyebrows, cheeks, and chin, and someone also appears to have drawn wings on the sides of the skull.[19] All of the examples I raise here, whether human skulls or the medical documentation of disabled people's institutionalized lives, illustrate the profound ethical risks of the commercialization and commodification of histories of medicine and the ethical quagmires that open up for the historian seeking to build a collection.

Aside from the obvious issues with the sale of human remains, I have also been thinking about what it means for my pedagogy. If I continued to use eBay, how would I prepare a student working on an exhibit in my class to encounter a human skull displayed in an antique shop? What happens when students confront entries on eBay like this? I do not claim to have all the answers here, but I have found that when I use the histories of medicine, empire, race, and disability to flesh out an ethical framework for students to consider such moments and incidents, it does help them understand the violence and genealogies of the processes of collecting and archiving. One of the central threads of my teaching is to describe how the disabled and racialized "Other" was routinely and regularly put on display, both dead and alive, for profit and for "science." When I teach, I make sure to humanize these stories and to center their agency and whatever traces of voice or experience I can find. In addition to the much more widely known story of Sara Baartman, one example I often use is that of the tattooed skin of "Prince Jeoly," the Mindanaoan man who was enslaved by William Dampier and then put on display until he died of smallpox, at which point his tattooed skin was preserved and displayed. Students are encouraged to understand the multiple ways in which empire resulted in his death, but also the indignities of display after his death. How did empire and its attendant violence, war, and desecration all play a role in facilitating these collections, and how did the museum carry on the violence of imperialism? These are the questions I want students to ask about all corporeal traces and other objects they find for sale in antique stores, in secondhand stores, on TikTok, or on eBay.

This chapter has examined ethical issues that arise when historians of medicine engage with, buy, or use objects or documents from online, commercialized spaces like eBay, which are in many ways important sites that provide access to popular historical narratives as well as sources that enable the work of precarious, underfunded scholars. Personally, I do not think I will ever be entirely "easy" with collecting. I am always uncomfortably aware of both the privileges necessary for collecting and the fact that collecting requires my participation in the commercialization and commodification of history. I sometimes also find myself thinking that my relationship with and use of my collections are different from those of the traditional collector, whom Walter Benjamin described as having "a relationship to objects that does not emphasize their functional, utilitarian value— that is their usefulness."[20] Additionally, I began collecting as a way to study histories that would otherwise require me to travel across oceans, navigate punitive visa regimes, and contend with my own chronic fatigue and illness. I would argue that collecting *can* constitute a partial methodological answer to the question of how to write the histories of people whose lives leave behind little trace in the archive or leave evidence that itself is "harmful, violent or incomplete."[21] But even so, how do we contend with the realities that collecting itself contributes to the "incomplete" archive? My solution is imperfect, but I present it here nonetheless: I am working with a couple of libraries and archives to donate my personal collections of various objects, and I ultimately also want to digitize and share my collections with a wider audience through open-access publications. At the same time, critically examining what collecting more generally means also gives us the chance to analyze the growing privatization and commercialization of our sources, and what it means to work in a space where the enfreakment and consumption of death, disability, and suffering are a business, and a profitable one at that.

NOTES

1. Walter Benjamin, *The Arcades Project,* trans. Howard Eiland and Kevin McLaughlin (Cambridge, MA: Harvard University Press, 1999), 883; Walter Benjamin, "Unpacking My Library," in *Illuminations: Essays, and Reflections,* ed. Hannah Arendt, trans. Harry Zohn (New York: Schocken, 1968), 59–67; Anne Pfeifer, "A Collector in a Collectivist State," *New German Critique* 45, no. 1 (2018): 49–78.

2. A very select list of such work includes E. P. Hamm, "Unpacking Goethe's Collections: The Public and the Private in Natural-Historical Collecting," *British Journal for the History of Science* 34, no. 3 (2001): 275–300; John Potvin and Alla Myzelev, eds., *Material Cultures, 1740–1920: The Meanings and Pleasures of Collecting* (Aldershot, UK: Ashgate, 2009); Joan Faber McAlister, "Collecting the Gaze: Memory, Agency, and Kinship in the Women's Jail Museum, Johannesburg," *Women's Studies in Communication* 36, no. 1 (2013): 1–27; E. Kuecker and M. Freeman, "The Aura of the Trace in One Child's Projects in the World: Collecting as Rescue, Repetition, Rupture and Refrain," *Journal of Childhood Studies* 46, no. 4 (2021): 32–45; and Anne Pfeifer, *To the Collector Belong the Spoils* (Ithaca, NY: Cornell University Press, 2023), 83.

3. See, for example, Vera Keller, Anna Marie Roos, and Elizabeth Yale, eds., *Archival Afterlives: Life, Death, and Knowledge-Making in Early Modern British Scientific and Medical Archives* (Leiden: Brill, 2018).

4. C. S. Lakshmi and Sruti Bala, "No More Sewing Machines! The Challenges of a Women's Archive in India," in *Gender and Archiving: Past, Present and Future,* ed. Noortje Willems and Saskia Bullman (Amsterdam: Uitgeverij Verloren, 2017), 103–107.

5. Craig M. Loftin, "Secrets in Boxes: The Historian as Archivist," in *Art and Queer Culture,* ed. Catherine Lord and Richard Meyer (London: Phaidon, 2013), 27–43.

6. Franklin A. Robinson Jr., "Queering the Archive," *QED* 1, no. 2 (Summer 2014): 195.

7. Gracen Brilmyer, "'They Weren't Necessarily Designed with Lived Experiences of Disability in Mind': The Affect of Archival In/accessibility and 'Emotionally Expensive' Spatial Un/belonging," *Archivaria* 94 (December 2022): 120–153.

8. Dydia DeLyser, "Collecting, Kitsch, and the Intimate Geographies of Social Memory: A Story of Archival Autoethnography," *Transactions of the Institute of British Geographers* 40, no. 2 (2015): 209.

9. DeLyser.

10. Zoe Trodd, "Hidden Stories, Subjective Stories, and a People's History of the Archive," in *Everyday eBay: Culture, Collecting and Desire,* ed. Ken Hillis, Michael Petit, and Nathan Scott Epley (New York: Routledge, 2006), 78.

11. Cecily Devereux, "Finding Indian Maidens on eBay: Tales of the Alternative Archive (and More Tales of White Commodity Culture)," in *Basements and Attics, Closets and Cyberspace: Explorations in Canadian Women's Archives*, ed. Linda M. Morra and Jessica Schagerl (Waterloo, ON: Wilfred Laurier University Press, 2012), 29–47.

12. Gracen M. Brilmyer, "Toward a Crip Provenance: Centering Disability in Archives through Its Absence," *Journal of Contemporary Archival Studies* 9 (2022): 12, https://elischolar.library.yale.edu/jcas/vol9/iss1/3/.

13. Ken Hillis, "A Space for the Trace: Memorable eBay and Narrative Effect," *Space and Culture* 9, no. 2 (May 2006): 140–156.

14. Aparna Nair and Kylie Smith, "We're Historians of Disability. What We Just Found on eBay Horrified Us," *Slate*, July 21, 2022, https://slate.com/technology/2022/07/vintage-asylum-records-found-on-eBay-history-of-disability.html.

15. Martha Mendoza, "Online Auction Hosts Trying to Limit Strange, Illegal Items," *Journal News* (White Plains, NY) September 11, 1999, 29.

16. James Hamilton, "Scottish Link to Child Remains on eBay History," *Sunday Herald* (Glasgow), November 5, 2006, 20; Sally Agpar, "Charges Filed in eBay Skull Case," *Honolulu Star-Bulletin*, September 9, 2004, 3; "EBay Knocks Skull's Auction on the Head," *Sydney Morning Herald*, May 25, 2000, 35.

17. Ryan M. Seidemann, Christopher M. Stojanowski, and Frederick J. Rich, "The Identification of a Human Skull Recovered from an eBay Sale," *Journal of Forensic Sciences* 54, no. 6 (November 2009): 1247–1253; Conor Gearin, "Hundreds of Mystery Human Skulls Sold on eBay for Up to $5500," *New Scientist*, July 12, 2016.

18. Seidemann, Stojanowski, and Rich, "Identification of a Human"; Gearin, "Hundreds of Mystery."

19. The eBay seller's page has been archived on the Wayback Machine, https://web.archive.org/web/20230811012705, site no longer available.

20. Benjamin, "Unpacking My Library," 62.

21. Brilmyer, "Toward a Crip Provenance," 4.

PART III
Research

CHAPTER TWELVE

Ethics and Ghosts beyond the Institutional Review Board

MARCO ANTONIO RAMOS

When the institutional review board (IRB) refused to review the ethics of my dissertation research on mental health amid the violence of the last dictatorship in Argentina (1976–1983), I was concerned. I called my IRB administrator to figure out what had happened. "Your project is exempt from IRB review because it's not research," she told me plainly. "Not research?" I managed after a few seconds. My project *was* research—I insisted—in which I recovered narratives of political resistance and repression. Ethical danger abounded. I planned to interview people who suffered from human rights violations, including political disappearance and torture. But she waved me away: "None of that matters." Citing section 46.102 of the Policy for the Protection of Human Research Subjects, she explained that the IRB defines research as "a systematic investigation . . . designed to develop or contribute to *generalizable* knowledge."[1] My project did not meet this criterion because—like most history—it was focused on a particular time and place. "But epistemology aside," I protested, "you must agree that I am producing *some* sort of knowledge with ethical implications." Annoyed by our dragging conversation, she ended it: "Again, your project is not research, so the question of *research ethics* does not apply." The IRB's positivist understanding of what counts as "research" exempted my dissertation research from ethical review.

This so-called oral history exemption has been well documented on the websites of universities and professional societies that offer helpful how-to advice for historians.[2] To my knowledge, however, these exemptions have rarely been interrogated with any sustained critical rigor. In fact, many historians and journalists have applauded these exemptions for "easing" the burden on their research.[3] As one friend told me, "Your exemption is a windfall—don't you know how frustrating the red tape of an IRB review committee can be?"

But I didn't feel relief—instead, I was concerned. I had resisted the IRB's exemption of my research because the ethical stakes of my project felt overwhelming. During preliminary trips to Argentina, I had learned that much of the archival material that I had hoped would serve as evidence for my dissertation had been destroyed during the dictatorship of General Jorge Rafael Videla. As one activist told me, she and her comrades were forced to burn, bury in backyards, or dump in trash cans the material traces of their activism in the 1970s for fear that their documents would become "passports to torture and death."[4] The military treated beloved books, meeting minutes, and diaries as evidence of "Communist subversion" that led to the political disappearance of thousands of citizens through secret imprisonment, torture, and murder.

The absence of archival material meant that my research would lean heavily on oral history. I worried that interviews with elder activists would dig up trauma that might cause more harm than good—friends disappeared, experiences of torture, the feeling of everyday repression that Argentine writer Ricardo Piglia described as "artificial respiration."[5] I had hoped that the arbiter of ethical research, the IRB, would help me manage this ethical complexity. I thought of its red tape, forms, and committees as blankets that would protect me from the ethical sharp edges of my research.

As the administrator had made clear, however, my project was exempt. My "not-research" existed in what Italian philosopher Giorgio Agamben calls "a state of exception."[6] In *Homo Sacer*, Agamben argues that the sovereign's political power resides in the ability to decide whether cases like mine are normal or exceptional. For Agamben, real power is found not in the application of the law itself but rather in the meta-power to decide whether the law applies at all. The IRB's sovereignty over research ethics was not simply reflected in the arbitration of "normal research" in committees, forms, and bureaucracies through all its red tape. Instead, its sovereignty resided in the power to define what was *not* research, through the creation of the exceptional zone in which I found myself. My IRB administrator flexed this sovereignty when she told me definitively, "Your project is not research, so the question of *research ethics* does not apply."

The zone of exception has not been addressed in most of the critical literature on the IRB. Scholars have tended to focus on the red tape that governs the ethics of normal biomedical science. Historians Nancy Campbell and Laura Stark, for example, have helpfully demonstrated that the IRB emerged around a legalistic vision of what it meant to be human in the 1960s. The narrow vision of the "vulnerable human" that animated the IRB was not only designed to protect research *subjects.*[7] Rather, the IRB functions also—and perhaps primarily—to protect research *institutions* from legal liability. After all, it is the university that pays the bills. The IRB's red tape was a cover for my institution's proverbial ass, legally speaking. But it would not serve as cover for me.

What, then, does ethics look like in the space of institutional exemption, in this ethical exile beyond the IRB? And what can this zone of exemption tell us about ethics and the university more generally? Agamben calls what happened to me "an abandonment."[8] His sense of abandonment resonates with me because the IRB's declaration that "research ethics does not apply"—quite obviously—did not mean that there were no research ethics here. Instead, it meant that the IRB would not help me work through my project's complicated ethics. I was cast into a space beyond the protection of the ethical sovereign and its stamps of approval. I was abandoned, "non-research."

I carry two intentions with me in this space of ethical exemption. First, I am writing to those who have found (and will find) themselves, like me, abandoned by the IRB. This essay is inspired by anxious conversations with other graduate students cast into the zone of ethical exemption. I write it for historians of science and medicine who are interested in what the violence of the past means for our understandings of health, the body, and perhaps most important, ourselves—especially those of us with personal ties to the historical harm that we study. But this is not a how-to guide for what to do when the IRB exempts your project (there are already plenty of those). Instead, please approach this essay as if it were a trip report: field notes from the ethical life beyond the IRB.

My second intention is to convey that being abandoned by the IRB is not the same as being alone. There is profound scholarship from the university's margins—from Black, Indigenous, and queer thinkers, scholars of the borderland, and those in disability studies—on ethics and the university that critiques and diverges from the narrow conceptions of humanity that undergird the IRB's work.[9] This scholarship is not only useful for those exempted from IRB review. Rather, following Agamben, I believe this work demonstrates how we have all been abandoned. Or to phrase it differently, it exposes how *the IRB has abandoned any deep sense of ethics that would make the university accountable to those whom it studies or claims to serve.* We are all abandoned by our institutions to struggle with the deeper ethics of our work—to grapple with the fraught histories of our fields, the lived impact of studying trauma, the unequal flow of resources in the university, the privileging of certain ways of knowing. Turned over to the IRB, we are left bereft by it.

It would be a mistake, then, to read this essay as an injunction for the IRB to include, instead of exempt, projects like my own. A policy change would not help here. The IRB's work centers on the "vulnerable human," and as I learned in Argentina, my work traffics in the ghosts of the Cold War. Here, I sketch the contours of the ethics that I developed to confront the ghosts from my family's past that have disrupted the neat boundaries between the personal and the professional, the human and the inhuman, and the past and the present that undergird traditional historical research. These ghosts have pushed me beyond the narrow ethics of the IRB, to explore how our field might serve as a means to care for intergenerational wounds, by reconfiguring our relationship to the violence of the past.

***

When I arrived in Argentina after the IRB exemption, I was eager to speak with elder activists who had lived during the dictatorship, especially Emiliano Galende. A psychoanalyst who escaped Argentina for his native Spain during the dictatorship, Emiliano was secretary general of the largest openly Marxist psychiatric organization in the country in the 1970s, the Federación Argentina de Psiquiatras.[10] In 2014, we engaged in a series of conversations about his psychiatric militancy in the 1970s. In our fourth conversation, however, he abruptly stopped talking about his past, looked at me squarely, and asked, "Marco, why are you so interested in this history?"

Concerned by the apparent suspicion that I had unintentionally provoked, I reassured myself that I was prepared for this question. I had won two prestigious grants that used (what I thought were) smart arguments about why this history was important. I offered Emiliano answers that riffed on familiar academic themes: the Latin American Cold War, the history of health activism, and science and technology studies in the Global South.

But none of these answers struck the right chord. Emiliano kept probing me with questions. Finally, I went off-script and started to tell him about my family. My father was a medical student activist under a different military dictatorship during the Cold War in Latin America. In Peru, in the 1960s, he was president of a medical student organization that vocally protested the authoritarian regime. After giving a rousing speech against the coup at the University of San Marcos, he was disappeared by military officers. For six months, he was transported between prisons to hide his location, as often happened with *desaparecidos* (the disappeared). Eventually, he escaped to Chile, and then to the United States, where I was born.

Those of you who live with intergenerational trauma might appreciate how my father's past was simultaneously common knowledge in my family but also rarely discussed explicitly. During my childhood, it felt like naming the obvious fact of my father's disappearance would allow the terror from the past to overwhelm and destroy us. Nevertheless, his narrative of disappearance and escape was ever present. It was, after all, the reason I was born in the United States and, perhaps, the reason he gave me his name—Marco Antonio Ramos—as proof that he still existed after he had almost disappeared forever. Much later, elder activists would give me the language to describe my situation more precisely. The unresolved violence of my father's past haunted my family as a ghost, a "presence-absence, an existence-nonexistence," that we inherited from the Cold War.[11]

As I rambled about my father to Emiliano, he suddenly interrupted me with a loud slap to his thigh. "Oh, now I see!" he exclaimed. "Your research is about your father." Bizarre as it seems now, this obvious fact had never occurred to me. I was not at a point where I was able to confront my family's ghosts so directly, and my training in the history of science and medicine had taught me to hide this repressed knowledge beneath sharp rhetorical moves and cutting-edge theory. Amused and embarrassed (of course a psychoanalyst thinks it is about my father—and of course he is right), I lied. I pretended that I had known about the deep connection between my father and research all along—a half-hearted deception that, with hindsight, was all too transparent.

Nevertheless, my relationship with Emiliano shifted with that self-revelation. In what must have been our eighth conversation, he told me, "Esperá un momento." He disappeared into his basement and reappeared a few minutes later with a couple of well-worn leather duffel bags. Inside the bags were the scattered and stained documents that constitute the entire archive of the Federación Argentina de Psiquiatras.

This gift changed how I understood ethical accountability in my research. Not only was I responsible for the consequences of digging into this traumatic history, but I also, quite literally, needed to be accounted for by my interlocutors. I had to be made sense of and oriented to. Emiliano wanted to know about my ancestors, my politics, why this history mattered across all the registers of my being. I assumed that I would be uncovering historical trauma from the period, but it was mine that would be uncovered because it was the same trauma, after all. I was in this history; it was in my life.

My conversations with Emiliano helped me appreciate just how alienated I had become from my work. Graduate training had carved me into distinct selves—psychiatrist, historian, Peruvian American, son of a torture survivor—and buried the connections between them in my unconscious. There was no room for my family's ghosts in graduate school.[12] As bell hooks argues, the traditional classroom functions as a space for theories and frameworks, not emotions and the body.[13] My study of Cold War history, as a professionalized field, obscured how this history haunted me. Graduate school pushed me toward a more perfect disembodiment that promised a certain authority and objectivity, but that required exorcizing the Cold War ghosts from my family's past.

My conversations with Emiliano, then, were acts of care that integrated my split subjectivity through a process that Frantz Fanon describes as "disalienation."[14] Refusing to respect the boundaries between my identities, the ghosts from my family's past emerged from the silence where they lay hidden and demanded a fuller integration of my selves.[15] For better or worse, my writing this essay is a response to these ghosts' demands, as I connect my research with the repressed knowledge and emotions that brought me here in the first place.

***

In his novel *Hades, Argentina*, Argentine American writer Daniel Loedel tells a ghost story, set in 2020, in which the protagonist, Tomás

Orilla, travels from New York City to his birthplace in Argentina to confront his trauma from the dictatorship. In an interview, Loedel discussed how the real-life political disappearance of his half sister affected his fiction: "There was . . . a ghost in my house growing up, a presence in the background. . . . In writing the book, I had an epiphany . . . that my emotional experience was one in which the ghost of this person haunted me—and haunted my father. . . . The idea [for the novel] was really less a structural decision than a means of accessing what is emotionally true about living in the aftermath of a disappeared loved one."[16]

Like Tomás, then, I was pulled to Hades, Argentina, to reckon with the "emotional truth" of living with the ghosts from my father's past. Perhaps unsurprisingly, most academic writing has failed to take ghosts seriously.[17] You might even think I am using *ghosts* here figuratively or because it is provocative. But "believe me, I believe in ghosts."[18] My experience in this zone of ethical exception has taught me that you are not communing with ghosts unless you feel them with an intimacy that can be both exhilarating and frightening.[19]

For example, just last year, I was discussing my project with my father, when he asked me to wait a moment. Like Emiliano, he disappeared into his closet and returned with a shoebox that was filled with the material traces of his activism from the Cold War. This secret archive of newspaper clippings and speech drafts has preserved for decades a life of political engagement that was repressed both politically, with his disappearance, and psychologically, as he relegated his militancy to an unmentionable past when he moved to the United States. For the first time, I am now interviewing him about this history, as I turn my constrained dissertation into a more undisciplined book on the afterlives of justice from the Latin American Cold War.

The editors of this volume asked me to write a "takeaway" from this zone of ethical exception. My point here is not simply that memoir or autobiography has ethical or academic value. Instead, I want to argue that historians of science and medicine should take the conditions and politics of their own research seriously. If we believe that the politics of knowledge are important, then our examination of those politics should begin with our own knowledge production. We should interrogate our motivations with as much rigor and curiosity as we treat those of our historical actors. Why do we study what we study?

And why at this point in our lives? Like it or not, we are all imbricated in the ever-unfolding history of the university that we strive so hard to critique. As experts of the past, historians should be better at reckoning with our own historicity.

To engage in this ethical reflexivity, we need to create more space for exploring our subjectivity—space to theorize our connection to the harm that we study, to feel the reverberation of trauma across generations.[20] Not unlike the IRB, the history of science and medicine retains a latent positivity in its methods and teaching that can erase our positionality and personhood. As Kathryn McKittrick and Linda Peake argue, professional engagement with capital-*H* History, as a (mostly white) field, requires many of us to abandon the lowercase-*h* history that we live and breathe.[21] Rather than alienating us from ourselves, how might our field serve as a means for making us whole, for making sense of the violence that is our inheritance? What if our ghosts were admissible in the classroom? If the IRB has abandoned us, then what kinds of communities and institutions can we create to reckon with the deeper ethics of our work?

I will conclude where this essay began: with a reflection on generalizability. Like my IRB administrator, a reviewer of this essay asked if my "ethical insights" were generalizable. In racially coded language, they wondered whether my insights only applied to researchers with "familial or generational trauma to draw upon" (as if my trauma is a privilege that benefits me unfairly). *What about the traumaless?* they wondered. This essay is for you too. I have offered my story as an example. But if you (with no trauma) do not know why you study topics like violence or harm, then you should also take an ethical look in the mirror. The university has a long history of exploiting the suffering of others, and no one stands outside history.

NOTES

1. Emphasis mine. "Information about IRBs and Oral History," Oral History Association, last updated July 2020, https://oralhistory.org /information-about-irbs/.

2. Historian Linda Shopes has explored the practical problems that oral historians face with IRB exemptions for the Oral History Association: Linda Shopes, "Oral History, Human Subjects, and Institutional Review Boards," Oral History Association, accessed November 18, 2024, https:// oralhistory.org/oral-history-and-irb-review/.

3. Colleen Flaherty, "Oral History No Longer Subject to IRB Approval," *Inside Higher Ed*, January 19, 2017, https://www.insidehighered.com /quicktakes/2017/01/20/oral-history-no-longer-subject-irb-approval.

4. Marcelo Brodsky, *The Wretched of the Earth*, photographic installation in Buenos Aires, 1995. See https://marcelobrodsky.com/en/nexus-8 -the-wretched-of-the-earth/.

5. Ricardo Piglia, *Respiración artificial* (Barcelona: Anagrama, 1980).

6. Giorgio Agamben, *Homo Sacer: Sovereign Power and Bare Life* (Stanford, CA: Stanford University Press, 1998).

7. Laura Stark, *Behind Closed Doors: IRBs and the Making of Ethical Research* (Chicago: University of Chicago Press, 2012); Nancy D. Campbell and Laura Stark, "Making Up 'Vulnerable' People: Human Subjects and the Subjective Experience of Medical Experiment," *Social History of Medicine* 28, no. 4 (2015): 825–848.

8. Also see João Biehl and Torben Eskerod, *Vita: Life in a Zone of Social Abandonment* (Berkeley: University of California Press, 2013).

9. This literature is too broad to cite here, but see, for example, Lily George, Juan Tauri, and Lindsey Te Ata o Tu MacDonald, eds., *Indigenous Research Ethics: Claiming Research Sovereignty beyond Deficit and the Colonial Legacy* (Bingley, UK: Emerald, 2020).

10. Marco A. Ramos, "Psychiatry, Authoritarianism, and Revolution: The Politics of Mental Illness during Military Dictatorships in Argentina, 1966–1983," *Bulletin of the History of Medicine* 87, no. 2 (2013): 250–278.

11. Diana Kordon and Lucila Edelman, "Efectos psicológicos de la represión política," in *Efectos psicológicos de la represión política*, ed. Diana Kordon and Lucila Edelman (Buenos Aires: XX, 1986), 27.

12. See Shannon Withycombe's essay in this volume.

13. bell hooks, *Teaching to Transgress: Education as the Practice of Freedom* (New York: Routledge, 1994).

14. Frantz Fanon, *Black Skin, White Masks* (New York: Grove, 2008). Also see Camille Robcis, *Disalienation: Politics, Philosophy, and Radical Psychiatry in Postwar France* (Chicago: University of Chicago Press, 2021).

15. See Ayah Nuriddin's essay in this volume.

16. Christopher Bollen, "Author Daniel Loedel on Ghosts, Tango, and 'The Odyssey,'" *Interview Magazine*, January 18, 2021, https://www .interviewmagazine.com/culture/author-daniel-loedel-explains-his-debut -novel-hades-argentina. Loedel's novel discussed in the interview is *Hades, Argentina* (New York: Riverhead Books, 2021).

17. Frederic Jameson, "Marx's Purloined Letter," in *Ghostly Demarcations: A Symposium on Jacques Derrida's Specters of Marx*, ed. Michael Sprinker (London: Verso, 2008), 39. In his chapter, Jameson argues that you do not need to "believe" in ghosts to appreciate their value for critical theory.

18. Jacques Derrida and Bernard Stiegler, *Echographies of Television: Filmed Interviews*, trans. Jennifer Bajorek (Cambridge: Polity Press, 2002 [original 1996]), 120.

19. See Super Futures Haunt Qollective, homepage, accessed November 18, 2024, https://superfutures.art; and especially Angie Morrill and Eve Tuck, "Before Dispossession, or Surviving It," *Liminalities* 12, no. 1 (2016): 1. Also see Katie Kilroy-Marac, "Speaking with Revenants: Haunting and the Ethnographic Enterprise," *Ethnography* 15, no. 2 (2014): 255–276.

20. Many essays in this volume make room for the kind of ethical self-reflection that I describe here. For example, see Jess Dillard-Wright's and Richard McKay's essays.

21. Kathryn McKittrick and Linda Peake, "What Difference Does Difference Make to Geography?," in *Questioning Geography: Fundamental Debates*, ed. Noel Castree, Alisdair Rogers, and Douglas Sherman (Malden, MA: Blackwell, 2005), 39–54.

CHAPTER THIRTEEN

Silences and Violence

AYAH NURIDDIN

Silences are manifestations of power that are produced by histories of racism and violence. The history of medicine needs a much more substantial engagement with the ways that archival silences are manifestations of power at multiple levels. Black studies, in its various instantiations, has long been concerned with questions of power that would be fruitful for historians of medicine to engage. Michel-Rolph Trouillot argues in *Silencing the Past* that examining power relations helps illuminate the process of historical production and that power shows us how silences enter the historical record.[1] Achille Mbembe describes the problem of "archivability," where only certain documents and materials are eligible for inclusion in archives, which also reflects the ways in which power operates in the past and present.[2] As historians of medicine, we might ask ourselves how we think about what has archivability. What has been excluded from the archives we use and why? How might we think critically and reevaluate what has archivability in our own work?

As historians of medicine, we need to examine the particular ways in which silences are produced in our field and our own work. How does silence affect the ways we write history? There has been inadequate recognition of the centrality of race and racism in histories of American medicine, which leads to the violent erasure of Black and Brown voices and experiences. Further, our field has not sufficiently dealt with how violence is embedded in the histories and practices of medicine and science. Violence has been treated as an aberration or as exceptional rather than central to American medicine and science. The very infrastructures of medical education, practice, and research have relied and continue to rely on violence and exploitation, and it is essential to confront this as we do our work.

In addition to documenting and analyzing violence in the past, historians of medicine also need to ensure that we do not commit what historian Marisa J. Fuentes calls "epistemic violence" by contributing to the continued erasure of Black voices and experiences from our work.[3] In this essay, I argue for using a Black power method as a framework to address the production of archival silences and amplify the voices and experiences of Black people. As articulated by N. D. B. Connolly, the Black power method calls on historians to challenge the definition of credible sources, to resist dominant white perspectives about the production of sources and archives, and to center Black knowledge and experiences.[4] It builds on the analysis of Trouillot, Mbembe, and other Black scholars to call for a fundamental rethinking of doing history, and it creates space to consider how this project could be applicable to other marginalized populations. It also serves as a call to think about the ethics of doing history. The Black power method can be useful and important for historians outside Black history because it makes visible the structures that silence marginalized populations in a variety of contexts. Ethical approaches to the histories of health and medicine must consider not only the ways that certain areas of content become silenced but also the methodological approaches that obscure marginalized and oppressed populations in the archive. This method helps historians to pose the question, How do we ethically write histories of violence and center those who experience harm?

This essay will use the Black power method to analyze archival silences in histories of American medicine at three levels of production: history, historiography, and professional space. It will focus largely on the ways that Black people and Black knowledge have been rendered silent in histories of American medicine. Yet these questions of erasure, silence, and violence are applicable to knowledge, voices, and experiences outside and beyond Black history. Engaging the Black power method in this way allows historians to do less harm by centering the voices and experiences of those most harmed by silences and violence both past and present.

History

At the level of history, Mbembe's concept of archivability shapes the kinds of sources that historians of medicine engage and consider essential to the field. What sources, voices, and experiences are centered in the history of medicine, and what is treated as peripheral? Until recently, much of the history of medicine did not sufficiently address the significance or centrality of race and racism. The scholarship that did this work often fell under the rubric of American or African American history even when explicitly dealing with questions of science and medicine. Black medical and scientific work was treated as separate from American medicine and relegated to an intellectual periphery. I use the term *intellectual periphery* to describe the ways that both historical actors and historians themselves have framed the work of Black people as tangential and insignificant, which has lasting consequences for how we write history. This framing has also obscured the ways that African American knowledge, clinical practice, and health advocacy are actually essential for studying the history of American medicine. Ethical approaches to histories of health and medicine require an examination of how intellectual peripheries are created and how to center the sources, voices, and experiences that were historically relegated to intellectual peripheries.

There are many examples of how African American knowledge, voices, and experiences have been obscured and even erased in histories of American medicine. One significant example from my own work is rooted in histories of African American public health work. In the early twentieth century, African American physicians, social scientists, public health workers, nurses, and others were particularly concerned about the state of African American health and beliefs that African Americans were a degenerating and dying race. African Americans dealt with disproportionately high infectious disease morbidity and mortality from conditions like tuberculosis, influenza, and venereal disease. They also dealt with the ongoing effects of structural racism: poor living and working conditions, a lack of access to adequate health care, and unrelenting discrimination. In response, African Americans developed eugenically inflected public health interventions to improve the health and composition of the race such as National Negro Health Week (NNHW) and other grassroots initiatives. NNHW and other Black public health efforts reflected that African Americans understood better health as part of a multifaceted struggle for racial justice.[5]

What is significant about Black public health, and NNHW in particular, is that African Americans were thinking broadly and critically about the relationship between health and the pursuit of equality. African Americans were arguing that racism itself, rather than prevailing notions of racial inferiority, was a public health problem. They had identified what we would now describe as social and behavioral determinants of health to challenge ideas of racial susceptibility to disease and assumptions of biological racial inferiority. African Americans were naming racism as a source of health inequality and thinking about health inequality in relation to political and social inequality in the same moments.[6] Histories of American public health often do not include or consider the significance of Black public health. The history of social determinants of health as a term and concept is often traced to the work of Sir Michael Marmot in the late 1990s and early 2000s, even though African Americans were wrestling with very similar and overlapping ideas nearly a century before.[7] Even though a number of scholars have written about African American health activism in its various forms, these works are often treated as peripheral to the larger histories of American public health and medicine. I would argue that this reflects the ways that African American history is framed as separate from and peripheral to larger movements in American history. Within the context of the history of medicine, it reflects the ways that African American voices, knowledge, and experiences continue to be relegated to the periphery and treated as separate from the central movements of histories of health and medicine. The aforementioned example is just one of the many moments of silencing and erasure of Black voices and experiences, and it is an opportunity to mobilize the Black power method to reevaluate how the field has been framed and constructed.

Historiography

At the level of historiography, Trouillot's framing of historical silences is particularly salient for analyzing how Black knowledge, voices, and experiences have been treated in histories of health and medicine. Much of the canonical work in the history of American medicine does not engage questions of race and racism. When it does, race and racism are treated as peripheral to the central questions of the history of medicine. Much of the canon represents a commitment to studying white intellectual production in medicine, and the canon continues to reify itself.

Though the canon itself is a subjective and fluctuating category, much of what would constitute a canon of American histories of health and medicine has centered what historian Antoine Johnson describes as the three-Ds perspective—doctors, drugs, and diseases—elsewhere in this volume. The canon has prioritized and centered white elite figures, white knowledge production, and white institutions. It has expanded because of social and cultural turns in the field, a growing investment in seeking patient voices and experiences, and an interest in looking at a broader set of clinical actors. Yet even with these important interventions, race and racism continue to be relegated to a corner of this historiography rather than treated as undeniably essential. In the historiography on American medicine, Black people and experiences are often discussed in two ways. First, they are often treated as victims and subjects of science and medicine. There are numerous examples of violence, victimization, and exploitation in the history of racist biomedicine.[8] From the assumptions embedded in ideas of racial susceptibility to disease to the various forms of experimentation conducted on Black people, this work highlights the relationship between science, medicine, and racial violence. It also emphasizes the ways that Black people continue to be targeted as medical subjects and objects, as well as how that process is interconnected with larger histories of American racism.[9] While writing about histories of violence and victimization is essential, a focus on violence and victimization often obscures the ways that Black people resisted that violence and created alternative discourses and visions for Black liberation. An emphasis on violence also runs the risk of relegating Black experiences to the realm of spectacle. As Saidiya Hartman shows in *Scenes of Subjection,* this emphasis not only can be reductive but in some ways also reproduces violence and harm.[10] Historians of medicine thus must consider how to document histories of racial violence ethically and effectively in medicine while not committing forms of epistemic violence in replicating harm or reducing Black experiences to moments of suffering and victimization.

Second, Black people become part of this historiography as exceptional practitioners in the histories of science and medicine. These narratives of superlative Black figures both counter and complement narratives of violence and victimization. These figures are often valorized for overcoming racism and other obstacles to become firsts in their respective fields. I have sometimes described this as the Black History Month approach to historiography, where Black figures are used to contribute to heroic narratives of achievement in the history of science and medicine. Biographies of Black physicians and scientists like Solomon Carter Fuller, considered the first Black psychiatrist, and biologist Ernest Everett Just are used to demonstrate the ways that Black people work past and beyond racism to make contributions to their field.[11] But focusing on valorizing certain Black figures produces another kind of silence. It causes us to imagine Black knowledge production and achievement as aberrations from a norm rather than as central features of the histories of American science and medicine. Though it is important to study the achievements of superlative Black figures, this kind of valorization can also obscure the impact of racism on Black knowledge production, Black institutions, and the ways that historians engage with their significance.

A "Black power method" approach to historiographies of American health and medicine would require pushing beyond the tendency to document violence or praise exemplary figures. Instead, it would require an examination of the centrality of race and racism to histories of American medicine. It would also necessitate more robust analysis and contextualization of contributions of Black people and a deeper engagement with the role of race and racism in the development of medical and scientific knowledge and institutions. Our understanding of histories of American science and medicine remains incomplete without a robust engagement with questions of race and racism.

The Profession

Silences in the historiography contribute to silences in the profession. Historians of American medicine and health need to consider the relationship between the writing of history and the composition of our field. Returning to Trouillot, it becomes necessary to think critically about power and the production of history. Who is writing the history of medicine? In which departments and institutions? What archives are used or not used? Which journals? Which professional societies? Who is present and who is deliberately made absent? Thinking through these questions creates space for historians of health medicine to examine the relationship between scholarship, the composition of the profession, and issues of advocacy. It also creates opportunities to imagine how a Black power method is useful for thinking about ethics in the historical profession.

The profession has moved toward greater and deeper engagement with questions of race and racism. The most recent catalysts for this work were the COVID-19 pandemic and the murder of George Floyd, the latter of which subsequently prompted a national conversation about racism and police brutality. In 2020, we saw an unprecedented number of academic institutions and societies begin to reckon with their own legacies of racism, such as through the creation of diversity committees and calls for renaming or removing problematic figures from institutions. Though this work is certainly important, it is also important to understand why it took a global pandemic and video recordings of Black people being lynched on the internet for our profession to meaningfully engage with these issues. It is also important to note that Black historians in various subfields have been raising these questions for a long time, and their concerns often have not been treated as serious or relevant to the historical profession. This too is another form of silencing. When our own colleagues are not believed about their lived experiences in the academy and in the world, it influences whether those experiences and voices are believed and taken seriously in scholarship.

Universities and professional societies are making important strides to address legacies of racism and inequality in the profession. There have been an increasing number of committees that focus on diversity, equity, and inclusion (DEI), as well as conversations and panels about what needs to change in the profession. But these DEI efforts have not made enough meaningful calls for structural transformation. This is not to say that DEI efforts in their various forms are not important or significant. Unfortunately, these institutional efforts are often constrained by what is palatable or legible to those institutions rather than reimagining what is possible in these settings. We collectively need to rethink what it means to be a part of this profession and think about the ways that many have been excluded or pushed out of the profession.

Rather than focusing on forming committees in which historically marginalized populations do the intellectual heavy lifting of rectifying inequality, we need to ask ourselves tough and uncomfortable questions about who is and is not included. Beyond looking at representational politics, we also need to think about breaking down barriers and transforming the field to ensure that as historians we do not contribute to ongoing forms of violence and inequality.

Conclusion

Addressing silences in our field at the levels of history, historiography, and professional spaces is essential for crafting ethical approaches to histories of health and medicine. This requires a deep and critical re-evaluation of how we do the work we do. What does it mean to put a Black power method into action? For many of us, this will require a deep look into the ways race and racism affect what it means to be a historian of medicine. It will also require an examination of the ways that other forms of inequality shape our field and how we engage with it. How do we think about the impact of sexism, ableism, Islamophobia, anti-Semitism, homophobia, and transphobia on our field? Which voices are amplified and which are silenced? This work will look different across our profession and various subfields, but all of us need to be asking these questions so that we can do less harm. Thinking about silence and/as violence requires us to engage capaciously and critically with scholarship and the contexts in which it is produced, as well as to imagine alternative possibilities for what our field could be.

NOTES

1. Michel-Rolph Trouillot, *Silencing the Past: Power and the Production of History* (Boston: Beacon, 1995), 26–27.

2. Achille Mbembe, "The Power of the Archive and Its Limits," in *Refiguring the Archive*, ed. Carolyn Hamilton et al. (Dordrecht, Netherlands: Kluwer Academic, 2002), 19–20.

3. Marisa J. Fuentes, *Dispossessed Lives: Enslaved Women, Violence, and the Archive* (Philadelphia: University of Pennsylvania Press, 2016), 5–6.

4. N. D. B. Connolly, "A Black Power Method," Public Books, June 15, 2016, https://www.publicbooks.org/a-black-power-method/.

5. Ayah Nuriddin, "Black Public Health," *Historical Studies of Natural Science* 51, no. 1 (February 2021): 151–154.

6. Nuriddin.

7. "Social and Behavioral Determinants of Health Data," Health Sciences Library Systems Guides, University of Pittsburgh, last updated November 18, 2024, https://hsls.libguides.com/social-behavioral -determinants-of-health-data.

8. Harriet Washington, *Medical Apartheid: The Dark History of Medical Experimentation on Black Americans from Colonial Times to the Present* (New York: Doubleday, 2006); James Jones, *Bad Blood: The Tuskegee Syphilis Experiment* (New York: Free Press, 1981); Alan Hornblum, *Acres of Skin: Human Experiments at Holmesburg Prison* (New York: Routledge, 1998); Jonathan Metzl, *The Protest Psychosis: How Schizophrenia Became a Black Disease* (Boston: Beacon, 2014); Edward Beardsley, *A History of Neglect: Health Care for Mill Workers in the Twentieth-Century South* (Knoxville: University of Tennessee Press, 1987).

9. Keith Wailoo, *Dying in the City of the Blues: Sickle Cell Anemia and the Politics of Race and Health* (Chapel Hill: University of North Carolina Press, 2001); Lundy Braun, *Breathing Race into the Machine: The Surprising Career of the Spirometer from Plantation to Genetics* (Minneapolis: University of Minnesota Press, 2014); Jonathan Kahn, *Race in a Bottle: The Story of BiDil and Racialized Medicine in a Post-genomic Age* (New York: Columbia University Press, 2013).

10. Saidiya V. Hartman, *Scenes of Subjection: Terror, Slavery, and Self-Making in Nineteenth-Century America* (New York: Oxford University Press, 1997), 3–4, 22–23.

11. Mary Kaplan, *Solomon Carter Fuller: Where My Caravan Has Rested* (Lanham, MD: University Press of America, 2005); Kenneth R. Manning, *Black Apollo of Science: The Life of Ernest Everett Just* (New York: Oxford University Press, 1984).

CHAPTER FOURTEEN

Patients and Survivors

JONATHAN SADOWSKY

Historians can land in sticky spots when we write about topics that are sensitive in the present.[1] We may be urged to tell our histories in certain ways and declare allegiances. Medical historians in recent decades have sought to capture the "patient's voice." But the "voice" can speak in different tongues. Some who get a medical treatment, for example, reject the term patient *and prefer the term* survivor. *Conflict over diction marks fraught terrain. As we traverse it, we face challenges over whose voices to honor and, in turn, what kind of story to tell. Here, I reflect on these challenges, using my own experience writing about the history of a controversial procedure, electroconvulsive therapy (ECT). I propose that narrative complexity can be an ethical calling.*

About twenty years ago, a stranger called me and asked to buy me coffee. I was writing a book about the history of ECT, often known as shock therapy, and he wanted to tell me about his ECT treatment. Historians of psychiatry can get nervous about the privacy of patients whose records we uncover. Do we have the right to see, let alone make public, records made with no intention of their use by nosy historians?[2] The anxiety is inflamed by stigma against mental illness. Here though, a psychiatric patient sought me out on purpose to get his story into the historical record. He wanted to counter not the stigma against mental illness but the stigma against a treatment for it.

At a Starbucks near my house, he told of the crushing depression he endured for decades. Despite a marriage and a successful career, he would spend his hours after work staring out the window, empty and futile. His gloomy withdrawal damaged his marriage. Psychotherapy and antidepressant drugs failed him. He finally decided to try ECT. People close to him urged against it, convinced it would damage his brain. After a short course of ECT, though, his depression cleared. A boulder he had borne for half a lifetime was lifted. He felt energized to engage the world. He had no adverse effects. The only downside, he said, was mourning the many years he had spent suffering when he could have been well. He wanted me to tell people what an effective treatment ECT is and fight its demonization. He gave me his life as data. Living proof.

He was the first, but others urged me to shape my narrative to conform with their own. I corresponded with an anti-ECT activist who called herself a survivor of the treatment. She had suffered permanent long-term memory loss following ECT treatments she had felt pressured into. Some short-term memory loss, especially of events near the treatment, is common with ECT, and often lost memories return. Permanent long-term losses are rarer but are distressing to patients.[3] She hoped I would tell people how barbaric ECT is and feared I would be fooled by psychiatric propaganda. I mentioned a famous ECT provider I had spoken with. She worried that he had won me over by giving me precious access to research materials. (I did ask that doctor for help with access to sources. He said no.)

When I gave public talks, an audience member usually objected strenuously to my account, from one side or the other—but I never knew in advance which side. Sometimes people quietly approached me to say they had been healed by ECT, with no serious adverse effects. One was offended by references, glib in her view, to literary sources that represent ECT. A relative of hers committed suicide after getting ECT, which she blamed on the ECT. She may have been right, but it is hard to be sure. ECT patients are usually people who have had severe depression for some time before the treatment. Ernest Hemingway took his life after an ECT treatment, but that followed years of mental decline that were likely due to massive substance abuse.[4] The woman's concern, though, showed the high stakes of representation.

A secondary goal of this chapter is to open discussion of aspects of the research process that we often omit in our formal discussions of method and theory.

The exchanges we have before and after talks, in hotel hallways during conferences, in casual emails with friends, and while teaching our students

*are rarely discussed in our published work. They may receive blanket ap-
preciations in the acknowledgments, but their specific roles in shaping the
printed product are obscured. Even the "personal communications" that occasionally populate our footnotes usually refer to formal correspondence
with experts. Informal exchanges exert force on our thinking, rivaling that
exerted by what we cite as our sources.*

*We often talk about bringing our research into our teaching. I have seen
this offered as a principal defense for doing research, with the desolate im-
plication that making knowledge only matters if it furthers pedagogy. But
aside from pious clichés about how much we learn from our students, we
talk less about the reverse: how we bring our teaching into our research. I
have tried in recent years to cite specific students on specific points when
they have helped me think about what I write.*

Clinicians weighed in, too. Psychiatrists often see dramatic improve-
ments in their patients after ECT. That is *their* experience. The fear felt
by potential patients frustrates them. "*One Flew over the Cuckoo's Nest*
came out in the 1970s!" After I gave a presentation based on a precir-
culated paper (the same paper that drew the accusation of glibness), a
doctor wrote me saying that my portrayal of ECT was too negative. He
raised a specific point he said my paper failed to address. I wrote back
and gave him the page number where I addressed exactly that point.
He wrote back and said, "Ah, I missed that." Another time I gave an
invited talk that was to be cosponsored by a history department and a
psychiatry department. When they saw the title of my presentation—
"Unsettling Stories of Shock Therapy"—the psychiatry department
withdrew its cosponsorship, saying that I was adding to the stigma
toward ECT. They had not seen the content of the paper, not even an
abstract. They might have been turned off by my use of the term *shock*,
which many psychiatrists think is misleading and adds to the negative
aura around the treatment.

Psychiatrists can be a little defensive about ECT.

Many of my interlocutors were invoking what Joan Scott famously
called "the evidence of experience," the powerful idea that the true experts are people who have lived something.[5] The ethos of "nothing
about us without us" has deserved power. But much great history writing would have to be thrown away if only people who had lived an

experience could write about it. The only history, ultimately, would be autobiography.

At a public function, my department chair once mentioned my work on ECT and said, "Jonathan intends to tell us that ECT isn't as bad as we think, though how would he know, since he's never had it?" I had never said whether I had or had not. I still do not know why he assumed I had not.

My project was not antiquarian, nor of interest solely to other historians. Most of the people I was studying were dead, but ECT is a living topic for many. For them, a lot rides on getting the history right.[6] The requests of people who want us to write a story that accords with their experience demand our attention, but we are not bound by them. But what if we were? What guidance do they provide when they disagree? As Scott stressed, it is hazardous to let any one experience stand in as *the* experience.

Academic colleagues also urged me to tell the story in certain ways. Rather than celebration or condemnation of ECT, many advocated neutrality, in the name of contextualization. Historians, anthropologists, and sociologists told me that our job is not advocacy for or against clinical procedures but the uncovering of context and meaning. I hope I am as much for careful reconstruction of context as the next historian, but this advice puzzled me, for two reasons. First, most historians, across subspecialties, now view the ideal of neutral analysis of context quaint to the point of naïveté. Advocacy is now more positively valued (a view I share). Second, assessing the efficacy of treatments is routine in medical history—as it should be. It is disingenuous to study procedures for years, take the trouble to master the science, and then say we have no opinions about efficacy. Historians of psychiatry are often cautiously agnostic about therapeutic benefits but abandon that modesty and restraint when assessing harm. Can we imagine writing a history of penicillin, chemotherapy for cancer, or the polio vaccine in which we bracket the question of whether they work? For my companion in Starbucks, the meaning of ECT *was* that it worked. The efficacy of a medical procedure cannot be severed from its context and meaning.[7]

I wonder if discomfort with evaluating ECT's efficacy showed a subtle form of ableism, in the form of a denial of real disability. ECT is a treatment for severe depression, and ECT is a strong form of medicine. Chemotherapy works, historians of cancer will say, but no one doubts that cancer is a sickness and a serious one. Did the discomfort with evaluating ECT therapeutically stem from mistrust of the identification of depression as a "real" illness?[8]

Another thing that rarely gets aired in our published works is the comments of peer reviewers—especially the negative ones! One reviewer of an article I wrote, long after the ECT book came out, said that my willingness to judge the efficacy of treatments showed that I did not understand the purpose of medical history. Oh well.

If the controversy over ECT pitted patients/survivors against providers, it would be easy to take a side. But the challenge of naming here is why it is not easy. In 1985, Roy Porter urged a medical history "from below," taking the "patient's view."[9] Porter intended a challenge to the doctor's point of view, even a form of resistance. Medicine, especially in the preceding 200 years, had become an "entrenched" and powerful authority. The patient's view, in Porter's manifesto, was the subaltern view in the history of medicine, analogous to the worker's standpoint in labor history. The call to "listen to patients" has inspired much rooting around in hospital records and other sources in search of documents to offer glimpses of patients' words and thus to challenge psychiatry. Few, if any, other medical specialties are routinely depicted by historians as so therapeutically barren, and worse, malevolent.[10]

Psychiatry offers much to criticize. It lacks diagnostic precision. It has a disturbing history of enforcing social conformity and hierarchy. Many of its treatments have neither a strong theoretical footing nor robust data for efficacy. Psychiatry offers few permanent cures. Often the best it can offer is symptom relief, partial and temporary. But many patients value psychiatry. If we wish to honor the voices of patients, we must honor those who value treatments, too. For people with severe mental illness, symptom relief is nothing to sneeze at.

ECT itself offers much to criticize, past and present. After over eighty years of use, we still lack a strong grasp of why it works. ECT was used abusively to control people confined in hospitals. Gay people were "treated" for the supposed illness of their sexuality. Even granting to the doctors the sincere belief that they were treating disease, this risky practice went on for years without a sliver of science showing that it brought the result the doctors sought, a change in sexuality. Eminent psychiatrist Donald Ewen Cameron used ECT as part of his ambitious and destructive program of psychic driving—dubbed, by Cameron himself, "annihilation therapy." Cameron's work destroyed minds and lives. And despite decades of research aiming to reduce adverse effects, ECT causes severe cognitive deficits and permanent memory losses in some patients. The science is painfully unclear about how common the worst outcomes are. *One Flew over the Cuckoo's Nest* continues to shadow perceptions of ECT, but clinicians who blame hostility to ECT on sensational media fail to reckon with the actual harms.[11]

And yet, from its inception, many have found ECT healing and even lifesaving. The majority of ECT treatments now are voluntary, and many recipients return for continuing treatment. Some patients find ECT healing, lifesaving, *and* devastating to memory and cognition. As we used to say on Facebook, it's complicated.

Some may deride this as "bothsidesism," as reflexively giving equal weight to two sides of an argument. Irresponsible or poorly supported views do not deserve equal weight. Historians should not seek balance for its own sake. We do not need to write the upside to the Tuskegee syphilis experiment because there was not one.[12] Writing the ambiguities in ECT history matters not because balance is a supreme value but because the history has real ambiguities. We may aspire to a reparatory history, but when those who have received a medical treatment hold such divergent views on its effects, what constitutes repair is not straightforward.[13]

Emotions run high in these debates. Given ECT's power to both relieve and cause suffering, it is natural for people to have strong feelings. I hope reasoned debate is possible anyway. To have that, we must see something simple to state but often surprisingly hard to grasp: experiences vary. Anti-ECT activists often ignore or dismiss the voices of patients who tell a positive story. ECT proponents are often equally guilty of dismissing patients with serious complaints as minority outliers. The latter dismissal sounds like an accusation of hysteria, with the disturbing, and famously gendered, valence that term evokes.

When my book came out, I hoped it might ease polarization in the debates over ECT. If Twitter—um, "X"—is any indication, it failed. Social media is now another influence on scholarly consciousness, and it is cited rarely. Social media is derided—with reason—as a poor platform for careful and reasoned debate. But a lot of debate does take place on it, and the debates on mental health are often predictably and lamentably polarized. The polarization is bad regarding psychiatric medications, but worse on ECT. Still, I have not lost hope that scholarly intervention might help turn debate from wholesale endorsements and condemnations toward more nuanced dialogue about particular costs and benefits. Maybe if Routledge marks down the price of my book enough . . .

Medical history matters in clinical encounters, if not always in narrowly instrumental ways. Doctors need to know that patient fear may not be irrational prejudice caused by scary movies but may stem from social or individual memory of real harms. Historiography conveys this more than clinical science. Patients contemplating ECT do need to know they are not entering *Cuckoo's Nest*. But clinical handbooks on ECT, and providers of ECT, often portray it as an essentially harmless, low-risk procedure. One prominent ECT proponent, quoting a patient, likened the therapy to a routine visit to the dentist. In the same book, he scorned any negative testimony by patients for tainting the reputation of a safe and effective treatment. He honored the voices of patients who told the story he wanted to tell, and dismissed those that were discrepant.[14]

Whatever our duty to the dead, as historians we have charges to the living, too. We fail people living with severe affective disorders if we render ECT solely as a form of social control or a remnant from a barbaric medical past. We should not deter them from treatments that could help. We also fail them if we depict treatments as risk-free. They are entitled to caution and foreknowledge. Anne Donahue suffered severe and lamented memory loss from ECT. She would, she said, have gotten the treatments anyway, had she been adequately warned. She had been told such losses were rare and felt "mocked by science."[15] The not being told hurt, perhaps as much as the memory loss itself.

When our projects prompt emotional debate in the present, we face pressures about how to render the past. Our most promising response is that topics are fraught precisely because their histories cannot be reduced to simple plot lines. An unambiguous argument can yield compelling narrative drive, persuasion, absorption, and for a lucky few, sales to a history book. When there is a relatively unambiguous argument to be made, it should be. But the wish for stories with a simple plot is hazardous.[16] Writing intricate stories does not only serve intellectual nuance. It can be an ethical obligation.

NOTES

1. This chapter has debts to many contributors to this volume, but I want to stress helpful comments by the editors and by Stephen Casper, Barron Lerner, and Sharrona Pearl. Also thanks to Damaris Puñales-Alpízar and Olivia Weiss for comments on a draft.

2. See Jonathan Sadowsky and Kylie Smith, "Reflections on the Use of Patient Records: Privacy, Ethics and Reparations in the History of Psychiatry," *Journal of the History of the Behavioral Sciences* 60, no. 1 (Winter 2024): e22260.

3. For a poignant account of permanent losses, see Anne B. Donahue, "Electroconvulsive Therapy and Memory Loss: A Personal Journey," *Journal of ECT* 16, no. 2 (2000) 133–143. I consider the rarity of permanent, long-term memory loss from ECT to be unproven. For my full discussion, see Jonathan Sadowsky, *Electroconvulsive Therapy in America: The Anatomy of a Medical Controversy* (New York: Routledge, 2016), esp. chap. 6.

4. See Kenneth S. Lynn, *Hemingway* (Cambridge, MA: Harvard University Press, 1987).

5. Joan Scott, "The Evidence of Experience," *Critical Inquiry* 17, no. 4 (Summer 1991): 773–797.

6. Rachel Louise Moran also describes hearing powerful but conflicting hopes and expectations from patients when she talked about her research on the history of postpartum depression. Rachel Louise Moran, "Spitting on My Sources: Depression, DNA, and the Ambivalent Historian," *Journal of the History of the Behavioral Sciences* 58, no. 4 (2022): 449–458. See also the chapters by Richard McKay, Marco Ramos, and Susan Reverby in this volume, which discuss the interest and stakes of the representation of the past to living people.

7. People who follow my work regularly (if such exist) have seen me make these points before. They are important to me (the points, I mean, but the people too). One might counter that the evidence for efficacy in psychiatry is weak compared with the interventions named earlier. Historians of psychiatry often underestimate the specialty's efficacy, but even if the evidence for it is weaker, that is part of the history, not a reason to skirt the question in favor of "meaning." The argument does not apply to ECT anyway, which has robust evidence for efficacy. Many opponents of ECT deny it, but it is true.

8. I examine the struggle of people with clinical depression to have their illness seen as "real" at length in Jonathan Sadowsky, *The Empire of Depression: A New History* (London: Polity Books, 2020), esp. chap. 6; and in an op-ed column, Jonathan Sadowsky, "Does John Fetterman's

Openness Signal New Acceptance of Mental Illness?," *Washington Post*, February 21, 2023. See also Moran, "Spitting on My Sources."

9. Roy Porter, "The Patient's View: Doing Medical History from Below," *Theory and Society* 14, no. 2 (March 1985): 175–198.

10. I will not give examples of the negativity in psychiatric historiography here, but I do in Jonathan Sadowsky, "Before and after Prozac: Psychiatry as Medicine and the Historiography of Depression," *Culture, Medicine, and Psychiatry* 45, no. 3 (June 2021): 479–502.

11. The themes in this paragraph are covered in Sadowsky, *Electroconvulsive Therapy in America*, chaps. 2, 3, and 6. Regarding Cameron, I have never seen a positive, grateful account of psychic driving, whether by a patient or someone close to a patient. We have, by contrast, many such accounts even of lobotomy. See Mical Raz, *The Lobotomy Letters: The Making of American Psychosurgery* (Rochester, NY: University of Rochester Press, 2015).

12. This is not to suggest that the story of the Tuskegee experiments lacks complexity; Susan Reverby's achievement has partly been to show how complex it is. Its complexity, though, lies in its overdetermination, not in any "upside." Susan Reverby, *Examining Tuskegee: The Infamous Syphilis Study and Its Legacy* (Chapel Hill: University of North Carolina Press, 2013).

13. On the concept and ideal of reparatory history, see Kylie Smith's essay in this volume.

14. Max Fink, *Electroshock: Restoring the Mind* (New York: Oxford University Press, 1999), 4.

15. Donahue, "Electroconvulsive Therapy," 138.

16. Richard McKay, a contributor to this volume, has eloquently shown how Randy Shilts's desire for a story with a clear villain did much injustice to the AIDS patient Gaëtan Dugas. Richard A. McKay, "'Patient Zero': The Absence of a Patient's View of the Early North American AIDS Epidemic," *Bulletin of the History of Medicine* 88, no. 1 (2014): 161–194.

CHAPTER FIFTEEN

Writing Ethical Medical History with Legal Sources

LAUREN MACIVOR THOMPSON

Navigating legal restrictions and using legal records in the writing of American medical histories is challenging. The 1996 Health Insurance Portability and Accountability Act (HIPAA) governs access not just for individuals' private medical information but also for historical material in the United States; archival medical records and documents may be redacted or even restricted completely. This volume's purpose is to broadly illustrate the ways that many historians and archivists have already led the effort in their thoughtful recommendations for balancing access and confidentiality in the use of historical medical records.[1] Historians also need to think critically about both the laws that govern medical recordkeeping and the use of legal records themselves as historical sources. This chapter, which is aimed at historians of the United States, explores best practices when examining legal records in the writing of American medical history. Yet the advice here is broadly applicable to those who work outside US history and can be read alongside the other essays in this volume on using a critical approach to sources, including those by Adria Imada, Michaela Clark, and Marco Ramos.

In the US context, HIPAA has functioned since its passage as an important safeguard against the exploitation of private patient information. Historians need to keep in mind, however, that the law can also perpetuate injustice long after the people it purports to protect are gone. Record restrictions, even ones that were established in good faith, can be another form of archival silence that reifies historical power differentials and shields institutions from liability in both the past and the present. As explored elsewhere in this volume, historians who by necessity are forced to work around this silence may unwittingly or even deliberately add to an existing narrative that was already concealing abuse and harm.[2] Legal restrictions on information that could otherwise bring long-delayed opportunities for restorative justice can be a form of revictimization for patients who were incarcerated or treated in the past for medical or psychological conditions. The delicate balance among historical subjects' right to privacy, the value of public disclosure, and the prevention of more harm should be a continual calculation when writing histories of medicine.

There are also important considerations about the ethical use of medical and legal records that are already available. Although HIPAA was modified a decade ago to specify that medical records could be made legally accessible fifty years after a patient's death, historical physician case notes, photographs, forms, and any number of other official documentations of an individual's experience with health, disability, and illness present important methodological and moral questions. Historians need to carefully parse these records to avoid making retroactive diagnoses or modern judgments about people's mental or physical capacities, and special attention should be paid to who produced them in the first place. Medical historians spend their time researching topics such as the historical development of medicines, vaccines, health technologies, and patient care approaches. To conclude that these advancements have improved human health and quality of life in the last few centuries is not incorrect, but it is only half the story. There remains the complexity of how "modern" medical advancements, including the advent of medical recordkeeping itself as a surveillance tool, was and continues to be a key source of legal, political, and social inequities.[3]

The same considerations apply to legal records, both current and historical. The researcher needs to consider the public nature of legal records in the first place. The law is our society's way to publicly negotiate justice and the consequences of an individual's actions, but as an apparatus of the state, its function is also social control. As legal historians of race and gender well know, law is not necessarily justice, nor is it made or enforced in a vacuum free of human prejudice. Instead, law has long shaped perceptions of human bodies and the accompanying degree of freedom one's body is afforded. A few perhaps obvious examples that are nevertheless worth reemphasizing are the facts that slavery was legal until 1865 in the United States and that American women were not allowed to vote until the twentieth century. Medical and scientific ideas about race and gender undergirded this construction of legal statecraft, defining exactly who deserved citizenship and protection by the Constitution.[4] Writing ethical histories of medicine therefore obliges us to explain more precisely the state's role in the shaping of individual health experiences both past and present. This is especially true when writing histories of medical or social reform institutions, which forces the researcher to interrogate the meanings behind official records.[5]

In my own work, I have faced these dilemmas with medical sources documenting miscarriage, abortion, and birth control, as well as case records that documented and judged my subjects' mental status and capacities.[6] In testimonies or depositions from physicians, and in judicial opinions, the voices of the people I study are often absent entirely. They are replaced instead with those of men in positions of power. As historian of disability Kim Nielsen has argued, court records that tell us about historical subjects' legal entanglements involving sex, marriage, divorce, custody, institutionalization, or guardianship often erase the individual in question and instead cause their existence, quoting Michel Foucault, to come down to only "exactly what was said about [them]."[7] In this vein, deposition records, testimony, and other documents included as evidence *can* be used to write ethical histories if we keep in mind that the archival evidence itself "shape[s] the meaning produced about them in their own time and our current historical practices."[8]

Medical historians who use legal records also need to be aware of the pitfalls of taking their conclusions at face value. Stephen Casper's excellent essay in this collection, titled "Expert Witnessing History," notes the gap between law and historians' evidence in the courtroom and examines the ways that historians serving as expert witnesses must work around the legal use of evidence, in contrast to their normal historical methodologies. As both Casper and I argue here, jurists and lawyers approach law and history differently. Their job is to select only the most favorable facts for the case, engaging in what the eminent constitutional scholar John Phillip Reid called "law office history." We need look no further than the latest Supreme Court decisions for robust examples of what happens when lawyers and judges employ history for their own ends.[9] Courts use historical precedent—relying on the principle of stare decisis—to make decisions, following the law as though it were somehow an entity that existed outside human concerns. Treating law and society as separate spheres neglects, however, the "fundamentally constitutive character of legal relations in social life."[10] Employing narrow histories and citing unexamined legal precedents obscures larger stories of state coercion that should be more carefully evaluated in the making of contemporary policy.

One salient example is examining the legal history of state and federal vaccine mandates. As the first doses of the vaccine for COVID-19 became available in early 2021, it was *Jacobson v. Massachusetts*, decided more than a century ago, that armed public health officials and employers with legal justifications for compulsory vaccination. In *Jacobson*, which stemmed from an era in which states and local jurisdictions were beginning to require smallpox vaccinations to attend school, the Supreme Court had declared that it was constitutional for states to enact mandatory vaccination laws and restrict individual liberties for the purposes of public health and the common good.[11] Yet jurists invoking *Jacobson* to either defend vaccine mandate requirements or justify the restriction of other constitutional rights during national emergencies have paid little attention to the decision's legal impact in its own time. Judge Oliver Wendell Holmes cited *Jacobson* in the infamous 1927 *Buck v. Bell* case, which declared involuntary sterilization constitutional in the name of public health. "Society can prevent those who are manifestly unfit from continuing their kind," he wrote. "The principle that sustains compulsory vaccination is broad enough to cover cutting the Fallopian tubes. Three generations of imbeciles are enough."[12] *Buck* has never been overturned, and, along with *Jacobson*, it is a reminder that contemporary decisions about public health rest on historical justifications for American institutions and the government to embed eugenics into active social policy.[13]

In the last half century, modern consent law has emerged as a powerful counterpoint to compulsory or involuntary medical research and procedures, but tracing the history of informed consent itself offers its own twists to reconsider. Most scholars point to the 1914 New York Court of Appeals case *Schloendorff v. New York Hospital* as a main origin point in the history of medical self-determination and patients' rights. The case involved a woman named Mary Schloendorff who sued the hospital after awakening from surgery to find her physician had performed a hysterectomy without her prior consent. She was being treated for stomach pain when her doctors also found a hard mass in her abdomen. They suggested that she be anesthetized so that they could better examine it, but Schloendorff decided that she wanted to return home instead of undergoing an exam. According to court records, the night before she was supposed to leave the hospital, nurses instead prepared Schloendorff for surgery despite her protestations and assured her it was merely so the surgeon could examine the tumor. Yet when she awakened from the procedure, she was in immense pain with a sutured wound across her abdomen and the inability to move her left hand. She then discovered the doctor had treated the tumor by removing her uterus, while the gangrene infection that ensued postsurgery eventually resulted in the loss of several of her fingers.

Schloendorff on its face is certainly worth teaching in bioethics, its blatant wrongs making a rational foundation for learning the parameters of informed consent based on Judge Benjamin Cardozo's opinion: "Every human being of adult years and sound mind has a right to determine what shall be done with his own body."[14] Researchers in other fields have also noted that the case can serve as a warning to contemporary physicians to maintain better communication with their patients, as "an enduring historical lesson . . . for modern gynecology."[15] Digging critically into the origins of a case like *Schloendorff* also offers other new perspectives, however. What is rarely mentioned is that Mary Schloendorff "lost both her uterus and her case," and most jurists subsequently interpreted the decision as establishing immunity for charity hospitals from the actions of their staff, not a declaration of patients' rights. As legal historian Paul Lombardo notes, "This was a very conservative opinion, yet one whose most memorable lines sound, in hindsight, expansively sympathetic to patients suing doctors."[16]

There are even more layers to the case when considering how the response of the physician defendants represented the deep entrenchment of medical paternalism in the early twentieth century. In Schloendorff's era, doctors had finally achieved the professional respect they had worked for in the prior century, and the source of public deference was their solemn, endlessly repeated pronouncements about the inferiority of female and nonwhite bodies in medical journals and public media alike. Cardozo's employment of the phrase "his own body" was beyond rhetorical flourish; rather, it reflected a specific understanding of *whose* body belonged to themselves by right of law. Schloendorff lost her legal suit because of attitudes in both law and medicine that deemed her and others as unworthy of that right.

Modern interpretations of *Schloendorff* as the origin of patient autonomy not only are anachronistic in terms of historical thinking but tend to frame the case as one that has somehow permanently fixed the problem of medical abuses. This is not to say that informed consent doctrine has not vastly elevated patient autonomy for many decades now.[17] As Susan Reverby has observed, "Horrific medical histories . . . are central [to] bioethical considerations." She writes, "In much of bioethics . . . cases of infamous wrongdoing play a crucial role" in galvanizing change.[18] But to couch *Schloendorff* as a triumph for patients' rights also belies the complexities and inequities for "informed consent" that remain in health care for many patients, including people of color, disabled people, trans people, women, and children.

Indeed, what happened to Schloendorff was, of course, merely one waypoint in the long history of racism and sexism at the heart of the American medical system. Other well-known medical exploitations followed her case, including the victimization of Black syphilis patients at Tuskegee in the 1930s and the cancer treatment of Henrietta Lacks, a Black woman patient at Johns Hopkins University in the early 1950s. In Lacks's case, Johns Hopkins was one of the few hospitals willing to treat patients of all races during segregation, but the physicians treating her for her cervical cancer also took samples of her cells for research without her knowledge or consent. Lacks passed away from the disease, and her samples (which became known as HeLa cells) have since provided one of the single most important biological foundations for the last seventy years of medical research and advancements. It is only recently, however, that Lacks herself has been honored for her contributions to science and medicine and that researchers have offered apologies to her family after exposing her medical information without consent or appropriate compensation.[19] The long-term ramifications of segregation and racism in American medicine continue to shape Black Americans' access to health care and health outcomes.

There is also the consideration of what "informed" might mean for some patient groups and the ways that it can actually perpetuate exclusion and harm. Transgender people now face a raft of state laws outlawing their medical care based on unproven declarations that doctors must inform them about their treatments because they are dangerous and permanently damaging to their bodies and psyches. Additionally, ideas about medical malpractice and informed consent have also often been wielded as a weapon in state legislative efforts to restrict and eliminate abortion access. "Crisis pregnancy centers" are an example of the ways that "informed consent" has collided with state coerciveness in forcing pregnancy for people seeking abortions.[20]

These examples illustrate why medical historians using legal sources should be wary of uncritically accepting official case decisions as neutral historical fact. We must be aware not just of archival silences and gaps but also the ways that both the law and the contents of the legal archive itself structure both the nature of our historical research and our conclusions. Keeping these various complexities in mind can have positive consequences for changing harmful narratives and obtaining justice. Good history exposes the truth, but the best history can be used to redress harm.[21] Excavating the limits and impacts of the law from and around our histories of science and medicine reveals not just the revelatory but the reparatory potential of our work.

NOTES

1. Judith A. Wiener and Anne T. Gilliland, "Balancing between Two Goods: Health Insurance Portability and Accountability Act and Ethical Compliancy Considerations for Privacy-Sensitive Materials in Health Sciences Archival and Historical Special Collections," *Journal of the Medical Library Association* 99, no. 1 (2011): 15–22. See also Aparna Nair and Kylie M. Smith, "We're Historians of Disability. What We Just Found on eBay Horrified Us," *Slate*, July 21, 2022, https://slate.com/technology/2022/07/vintage-asylum-records-found-on-ebay-history-of-disability.html.

2. Saidiya Hartman, "Venus in Two Acts," *Small Axe* 12, no. 2 (June 2014): 1–14; Michel-Rolph Trouillot, *Silencing the Past: Power and the Production of History* (Boston: Beacon, 1995).

3. "Roundtable: Historians and the Question of 'Modernity,'" *American Historical Review* 116, no. 3 (2011): 631–637. See also Michel Foucault, *The Birth of the Clinic: An Archaeology of Medical Perception* (New York: Vintage Books, 1973); Deirdre Cooper Owens, *Medical Bondage: Race, Gender, and the Origins of American Gynecology* (Athens: University of Georgia Press, 2017); Harriet Washington, *Medical Apartheid: The Dark History of Medical Experimentation on Black Americans from Colonial Times to Present* (New York: Doubleday, 2006); and Michael Willrich, *Pox: An American History* (New York: Penguin, 2011).

4. Hendrik J. Hartog, "Pigs and Positivism," *Wisconsin Law Review* 1985 (1985): 899–935; Rana A. Hogarth, *Medicalizing Blackness: Making Racial*

Difference in the Atlantic World, 1780–1840 (Chapel Hill: University of North Carolina Press, 2017); Nancy Isenberg, *Sex and Citizenship in Antebellum America* (Chapel Hill: University of North Carolina Press, 1998); Barbara Welke, *Recasting American Liberty: Gender, Race, Law, and the Railroad Revolution, 1865–1920* (Cambridge: Cambridge University Press, 2001); Christopher Willoughby, *Masters of Health: Racial Science and Slavery in U.S. Medical Schools* (Chapel Hill: University of North Carolina Press, 2022).

5. Model examples of this kind of work include Susan Burch, *Committed: Remembering Native Kinship in and beyond Institutions* (Chapel Hill: University of North Carolina Press, 2021); Wendy Kline, *Building a Better Race: Gender, Sexuality, and Eugenics from the Turn of the Century to the Baby Boom* (Berkeley: University of California Press, 2005); Rebecca Kluchin, *Fit to Be Tied: Sterilization and Reproductive Rights in America, 1950–1980* (New Brunswick, NJ: Rutgers University Press, 2011); and Natalie Lira, *Laboratory of Deficiency: Sterilization and Confinement in California, 1900–1950s* (Berkeley: University of California Press, 2021).

6. Lauren MacIvor Thompson, "'The Presence of a Monstrosity': Eugenics, Female Disability, and Obstetrical-Gynecologic Medicine in Late 19th-Century New York," *MIRANDA* 15 (November 2017), https://doi.org/10.4000/miranda.10536; Lauren MacIvor Thompson, "'The Reasonable (Wo)man': Physicians, Freedom of Contract, and Women's Rights, 1870–1930," *Law and History Review* 32, no. 4 (November 2018): 771–809.

7. Kim Nielsen, *Money, Marriage, and Madness: The Life of Anna Ott* (Urbana: University of Illinois Press, 2020), 7–8.

8. Marissa J. Fuentes, *Dispossessed Lives: Enslaved Women, Violence, and the Archive* (Philadelphia: University of Pennsylvania Press, 2016), 2.

9. John Phillip Reid, "Law and History," *Loyola of Los Angeles Law Review* 27 (1994): 193; Lauren MacIvor Thompson, "Held: Legal Authority and the Abuse of History," *Perspectives on History*, April 2023, 19–21.

10. Robert W. Gordon, "Critical Legal Histories," *Stanford Law Review* 36 (1984): 104.

11. Josh Blackman, "The Irrepressible Myth of *Jacobson v. Massachusetts*," *Buffalo Law Review* 70, no. 1 (2022): 131–270.

12. Jacobson v. Massachusetts, 197 U.S. 11 (1905); Buck v. Bell, 274 U.S. 200 (1927).

13. Edwin Black, *War against the Weak: Eugenics and America's Campaign to Create a Master Race* (Washington, DC: Dialog, 2012); Kline, *Building a Better Race*; Stefan Kühl, *The Nazi Connection: Eugenics, American Racism, and German National Socialism* (Oxford: Oxford University Press, 1994); Lira, *Laboratory of Deficiency*; Paul Lombardo, *Three Generations, No Imbeciles: Eugenics, the Supreme Court, and "Buck v. Bell"* (Baltimore: Johns Hopkins University Press, 2008); Alexandra Minna Stern, *Eugenic Nation: Faults and Frontiers of Better Breeding in Modern America* (Berkeley: University of California Press, 2016); James Q. Whitman, *Hitler's American Model: The United States and the Making of Nazi Race Law* (Princeton, NJ: Princeton University Press, 2017); Michael

Willrich, "The Two Percent Solution: Eugenic Jurisprudence and the Socialization of American Law, 1900–1930," *Law and History Review* 16, no. 1 (April 1998): 63–111.

14. Schloendorff v. New York Hospital, 211 N.Y. 125, 129–130 (1914). See also Tom L. Beauchamp and James F. Childress, *Principles of Biomedical Ethics* (Oxford: Oxford University Press, 2013); Frank A. Chervenak and Laurence B. McCullough, "Ethical Issues in Cesarean Delivery," *Best Practice and Research: Clinical Obstetrics and Gynaecology* 43 (August 2017): 68–75; Raymond J. Devettere, *Practical Decision Making in Health Care Ethics: Cases, Concepts, and the Virtue of Prudence* (Washington, DC: Georgetown University Press, 2016); Ruth R. Faden et al., *A History and Theory of Informed Consent* (Oxford: Oxford University Press, 1986); and "Informed Consent," in *Encyclopedia of Bioethics*, ed. W. T. Reich (New York: Simon and Schuyler Macmillan, 1995), 3:1234.

15. Judith Chervenak, Laurence B. McCullough, and Frank A. Chervenak, "Surgery without Consent or Miscommunication? A New Look at a Landmark Legal Case," *American Journal of Obstetrics and Gynecology* 212, no. 5 (May 2015): 589.

16. Paul A. Lombardo, "Phantom Tumors and Hysterical Women: Revising Our View of the Schloendorff Case," *Journal of Law, Medicine and Ethics* 33, no. 4 (2005): 794–795.

17. Other landmark informed consent cases include Mohr v. Williams, 95 Minn. 261, 104 N.W. 12 (1905); Salgo v. Leland Stanford Jr. University Board of Trustees, 154 Cal. App. 2d 564 (1957); and Natanson v. Kline, 186 Kan. 393 (1960).

18. Susan M. Reverby, "Ethical Failures and History Lessons: The U.S. Public Health Service Research Studies in Tuskegee and Guatemala," *Public Health Reviews* 34, no. 1 (June 2012): 13.

19. Dorothy Roberts, *Killing the Black Body: Race, Reproduction, and the Meaning of Liberty* (New York: Vintage Books, 1997); Rebecca Skloot, *The Immortal Life of Henrietta Lacks* (New York: Crown, 2010); Washington, *Medical Apartheid.*

20. Kassandra DiPietro, "Who Decides? Informed Consent Doctrine Applied to Denial of Reproductive Health Care Information at Crisis Pregnancy Centers," *Iowa Law Review* 107 (2022): 1253–1281.

21. See, for example, Presidential Commission for the Study of Bioethical Issues, *"Ethically Impossible": STD Research in Guatemala from 1946 to 1948* (Washington, DC, September 2011); Rachel Nostrant, "Vermont's Legislative Leaders Apologize for State-Sanctioned Eugenics Movement," *VTDigger*, October 18, 2021; and Linda Villarosa, "The Long Shadow of Eugenics in America," *New York Times*, June 8, 2022.

CHAPTER SIXTEEN

Expert Witnessing History

STEPHEN T. CASPER

It is the nineteenth day of trial—November 3, 2022. I have no legal expertise, but to me it looks like the plaintiffs are losing. The medical case before the jury is awful: Was Matthew Gee's death caused by playing football at the University of Southern California, or was it caused by his cocaine abuse, alcoholism, and numerous comorbidities? The plaintiffs argue that football-induced brain damage triggered a chain of events leading to his premature death.[1] The defendants claim that football does not cause brain damage. They argue Gee's death can be attributed to his addiction, obesity, and sleep apnea. It has been heart-wrenching to witness his children and wife endure this trial, reliving the debate through each expert's testimony. Equally disheartening is knowing that the outcome of this case may be twisted to bolster the illusion of football's safety.

When I first agreed to provide expert testimony for plaintiffs in a concussion class action called the *Matter of the NHL* in 2015, I worried about the ethics of such work.[2] Academics, on the surface, view the influence of money with suspicion, recognizing how it can manipulate judgment.[3] History has many scars from courtroom debates, particularly from the tobacco wars.[4] Historian Richard Evans, in his discussion on testifying about Holocaust denial, emphasizes that not all interpretations of history hold equal weight, and he suggests historians act as advocates for history.[5] Lauren MacIvor Thompson observes in this volume that "good history exposes the truth, but the best history can be used to redress harm." In a similar spirit, I consider myself an advocate for history, and the historical record is ultimately why I agree with plaintiffs in concussion cases.

With hindsight, I realize that my initial reservations, rooted in a narrow theory of professional objectivity, have evolved. Expert testimony, subject to real-time peer review, can be discredited if deemed unreliable by the court. Attorneys exploit opportunities to reveal bias. Defendants restrict their experts' access to documents, while plaintiffs rely on me to introduce historical evidence. To ensure a fair presentation of all available information, I endeavor to be encyclopedic. Despite my best efforts, the court may exclude evidence I rely on.

In this essay, I generalize from my professional experience to argue for a particularistic ethics of visibility.[6] I believe not testifying would cause harm. Translating historical knowledge into admissible evidence, relevant for juries and judges, enables their deliberative process. Visibility alone does not lead to agreement—if it did, history would be an easier way of knowing.[7] To address narrow legal questions, historians rely on a vast expanse of time, diverse primary sources, oral testimonies, contextual factors, secondary literature, and their expertise.

The goal is to assist the jury in making fair decisions in complex circumstances. Since I am not a lawyer, my understanding of how this works is limited, but I can confidently say historians play a critical role in permitting the past its hearing. Court rulings will ensure that the act of expertly witnessing history will be limited, and so historians must use their skill to communicate to the jury as completely as possible. As Susan M. Reverby makes clear in her profound conclusion to this volume, failing to do so may ensure that the past remains unwitnessed legally.

Expertise and Research

Historians have a seemingly insignificant "superpower." We pronounce evidence *from the past* evidence *of the past.* In our craft we masterfully study sources and interpretations of them to select among those representative examples of the broader chronology. The ability to judge evidence as a fact that would and should inform historical analysis is a cultivated intuition, learned by our training and extensive practice. Coupled with sworn testimony, this "superpower" permits the sources the historian relied on to become part of the evidence the jury consults.

Remember: it is the nineteenth day of trial—November 3, 2022. Days before, I testified for hours. I recalled hundreds of corporate records—both for the defendants and for the plaintiffs. I had been frustrated because the court excluded journalistic sources from my analysis. I have since left the Los Angeles district court. I am watching these proceedings on television.[8] William Horton, the plaintiffs' attorney, is cross-examining one of the star witnesses of the National Collegiate Athletics Association (NCAA), James Puffer.

Puffer is a team physician at the University of California, Los Angeles. To me, Puffer qualifies as an NCAA insider. He seems like a nice guy, albeit paternalistic. He has basically seen no records from the NCAA. His grandfatherly demeanor is probably reassuring for the jury. I explained in my testimony earlier in the trial that when Puffer was on the NCAA's Competitive Safeguards and Medical Aspects of Sports Committee in the 1980s, the NCAA did not have concussion guidelines. Gee had played from 1988 to 1992. Years before he played, a player had died from sequential concussions. Puffer's safeguards committee had pledged themselves to writing a concussion protocol. They never did.[9]

Horton is testing Puffer's present-day views. Horton says, "There's a lot of literature the jury has seen about the link between repetitive head trauma and long-term neurodegenerative disease. . . . And I know you disagree with it; fair?" Puffer responds, "Yes, I do." Horton continues, "Fine. Different people agree about different things. I'll accept that. But why is it that you get to decide whether or not I'm going to warn a student athlete about a condition that could leave him crippled or dead years later and not just share the information with the athlete and let them make their own decision?" Puffer answers, "Because I don't have good evidence that is in fact the case. And in fact, there's not good evidence—irrefutable evidence and literature to support that fact that is indeed the case."[10]

Horton spots that Puffer has been unreasonable: "So your qualifications for warning a student about a disease that could affect him or her 10, 20, 30 years down the road and leave them debilitated with brain injury is it has to be irrefutable and not just good evidence." Puffer rejoins, "Well, I think the standard of care is such that a reasonable physician, aware of the information that's available at that time, would make a decision as to whether it was appropriate or not appropriate." "What if you're wrong?" Horton asks. Puffer answers calmly, "Then I'm wrong."[11]

Horton appears baffled by Puffer's answer: "What are the consequences of being wrong, though, with a bunch of 18- and 22-year-old kids?" Puffer explains, "If I'm wrong, we would expect that there would be an epidemic of neurogenerative disease." Horton interjects, "There is." Puffer shoots back, "There's not. . . . Let's just say over the last 10 years, let's—you've got 250,000 people that have played intercollegiate football, and you would expect that if in fact this were a real phenomenon, that we would have lots and lots of people that have neurodegenerative disease as they get older."[12]

I find this exchange devastating. The exclusion of journalism weeks before means that the jury does not have access to a crucial document—evidence that Puffer has probably never seen either. A few years before, *Sports Illustrated* published an article about five linebackers who played in the same period for the University of Southern California, including Matthew Gee.[13] Four are dead. Gee is the fifth. The judge has excluded this evidence as hearsay; the exclusion of journalism ensures that this article will not shape the jury's views of Puffer's testimony. Puffer's testimony cannot be challenged by an alarming epidemiological fact.

Beyond this, I assume the putative reasonableness of Puffer's statement is shaping NCAA policy. Even as the National Institutes of Health have declared recently that exposure to repeated head impacts may lead to neurodegenerative disease, NCAA decision makers are unwilling to warn college athletes about that fact.[14] While their defense attorneys know about the *Sports Illustrated* article, likely none of the NCAA leadership are aware of it. In other words, the exclusion of evidence is both affecting this case and not forcing leaders to reckon with an unsettling reply from the past.

Testifying under Oath

It is day twelve in *Gee v. NCAA*, October 25, 2022. I am on the stand. To me, Bill Molinski, the defense attorney, looks a bit sheepish as he approaches me. Defense has decided to voir dire me. I am the first expert the jury is seeing. Minutes in, Molinski asks me, "You recalled you were deposed in this case?" I respond, "Yes." He says, "I'm going to show you a clip from that deposition for the record, to see if we can refresh your recollection." There is a dispute over whether this is proper impeachment. The jury is confused. I am confused. The judge allows the deposition tape to be played. To me it seems like the video of me says what I just said. William Horton, the plaintiffs' attorney, looks at me and snorts in derision at the theater. Molinski unabashedly moves forward. The whole exercise is intended to undermine my credibility and disorient the jury.[15]

When I first began studying the NCAA, I had little understanding of the organization and its role in American academic life. After years of studying its corporate records and reading its insiders' accounts, depositions of employees, and academic critiques of its influence, I have developed a hostile incredulity at its business model. I told the lawyers this during my deposition; I also tell the jury. But I am not allowed to tell them why. What you say in a deposition is rarely what you are allowed to say in a trial. Depositions tend to let everything in, and thereafter lawyers seek to exclude the experts and, if that fails, expert evidence.

Because deposition testimony is sworn, the expert can render their own testimony irrelevant to the litigation by simply not preparing appropriately. For example, in one case, I was asked to cite verbally every document that I considered salient between 1933 and 1960. I recited from memory dozens of sources that I knew were salient to the litigation. Had I been unable to recall these documents, then it is possible that the lawyer might have convinced a judge to limit my testimony to the facts only contained in those documents I recalled.

Because of the seriousness of deposition testimony, memory games are often not the only ones the expert encounters. Premeditated intimidation can be central in the experience of the expert. In one action in which I participated, I was ushered into a firm in Washington, DC, and our team was greeted by a senior partner who dragged me, as a historian, to a room overlooking the White House. "This is where Bill Clinton gave testimony. . . . You think you belong here?"

Back on day twelve in *Gee v. NCAA*, October 25, 2022. The battle taking shape before the jury is over experts. The defense needs to limit my frame and evidence; ironically they also need me to bring in their preferred evidence. This they will do over interminable hours during cross-examination, asking me, to the near death by boredom of each jury member, whether I relied on a document and the next and the next, to which I inevitably answer yes.

Historians occupy an unusual position in litigation because they can help plaintiffs and defendants understand how complex organizations behaved in the past and what individuals in those organizations did. Moreover, historians can also assess the context of those behaviors and actions and help everyone understand whether the actions seem outside the character of the times or appropriate to them. While I find the way courts exclude evidence baffling, I think historians should try to ensure as much evidence as possible is available to the jury and judge.

The Law's Rules of Evidence Are *Not* Historians' Rules of Evidence

It is now day twenty-eight—November 17, 2022. Defense expert Douglas Wiebe, professor of epidemiology and director of the University of Michigan Injury Prevention Center, is being cross-examined. Wiebe has been discussing a large study initiated by the NCAA and the Department of Defense to explore whether exposure to repetitive head impacts leads to neurodegenerative disease. Although Wiebe is not involved in the study, he argues that this study is the kind of work necessary for clarity. The plaintiffs' attorney asks, "So how long is it going to take before there's a consensus on repetitive head trauma and CTE [chronic traumatic encephalopathy]? Do you have a date in mind?" Wiebe responds, "No. How long it will take, you know, it's going to be some years. We see investigators here, myself included, saying what we need is a longitudinal prospective cohort study, pull together a group of athletes and give them time to age and get older, and eventually some day die. And it's that kind of study that will give us strong evidence and be above bar and able to test causation. So that's a while." The follow-up triggers an objection: "Fair amount of people will have to die of CTE if you're wrong on this; right?" The judge overrules. Wiebe says, "Time will pass and I don't know what number of people will die from CTE."[16]

Of course, the NCAA has more than a general knowledge about this question, but only it knows what it knows. I doubt Wiebe knows how many concussion cases the NCAA is battling; it has no reason to tell him. Likewise, Wiebe probably does not know that Senator Chris Murphy of Connecticut has relied on a plaintiffs' actuary report from another case to explain that thousands of college students who played football from 1958 to 2008 are expected to die.[17] Nor can Wiebe be asked about professional football players who are receiving payments through the unrestricted settlement from the National Football League. In this kind of abstracted legal universe, the past poses no ethical question.

Wiebe's testimony reflects the logic of progress. The past is gone, and when the past is excluded, the future is the only solution for the challenges of the present. Why have the contradictions to Wiebe not been brought forward? All of it is excluded. Mention of past or pending lawsuits is considered prejudicial. Settlements are considered non-admissions of liability. They, too, are excluded. Records from those settlements are considered hearsay. Even the fact that professional athletes have CTE is taken as evidence that college athletics may not cause disease. In many a courtroom, the rules of evidence limit all such historical context.

Among the most challenging aspects of expert witnessing history is that what counts as evidence in the courts is narrower than what counts for historical interpretation. Debates about evidence often endeavor to carve out advantages while limiting the ability of the expert to bring their full view of the matter into evidence before the jury.

It is months after Gee; the trial was lost. I am reading a rough transcript of a case fought in Indiana in January and February 2023. The case is *Finnerty v. NCAA*, and the chief medical officer of the NCAA, Brian Hainline, has been called to testify by the plaintiffs. The rough transcript is a public record, but it is not clear on what specific date this testimony occurred. I sat this case out. I am reading it now, wondering if it sheds new light on this story.

Is my absence making it tricky to bring evidence into the case? I do not know. Plaintiffs and defendants are in a heated argument over whether presentation slides I would have relied on are admissible. Moments before, the plaintiffs have said: "And it [the title slide] says 'Brian Hainline, NCAA Chief Medical Officer' on the front cover there?" Hainline responds, "Yes." On the surface, this should establish authorship. Yet what follows are pages of discussion including periodic questioning of the witness. The plaintiffs' lawyer states, "Your Honor, Indiana law is explicitly clear that a party may adopt another statement through conduct or otherwise. If the party has manifested an adoption or believe in the truth of the statement, the statement is not hearsay when offered against that party."[18]

An excerpt from the transcript of day eight of the trial captures the narrow argument that follows. The issue is whether the slides' content constitutes the content of the author's views or is instead an effort by plaintiffs to use the presentation to inject hearsay into court proceedings by another means. The court resolves the issue on a slide-by-slide basis.

184   ROUGH DRAFT TRANSCRIPT

1   explicitly clear that a party may adopt another
2   statement through conduct or otherwise. If the
3   party has manifested an adoption or believe in
4   the truth of the statement, the statement is not
5   hearsay when offered against that party.
6   Dr. Hainline to—
7   THE COURT: I know, but—I'm not
8   saying—I just need to know that these slides
9   you want to use, that he said these things.
10   That's all I need to know.
11   Because the problem is you're talking about
12   a PowerPoint and that rule, when it was drafted,
13   didn't talk about PowerPoints, right? So I'm
14   not saying you can't do it. I just need to make
15   sure that these are the things he told whoever
16   he was making the presentation to.
17   ██████: Your Honor, we have no
18   objection to Dr. Hainline's own words. He's
19   happy to talk about the presentation that he's
20   made, Your Honor. I think the fundamental
21   question is are the words on his own page or were
22   they taken from another source. The witness is
23   prepared to answer that question.

The level of argumentation that follows each slide is extraordinary. In a subsequent exchange with the chief medical officer, the plaintiffs' attorney endeavors to bring into evidence a famous article from 1928 that I also would have relied on. The screenshot of the article appears in the PowerPoint, and the plaintiff tries to establish that a screenshot means it is within the witness's expertise.

15   A Yes.
16   ██████: Okay. Your Honor, I would move
17   to admit P12 into evidence.
18   ██████: Objection. It's hearsay,
19   Your Honor.
20   ██████: It's ancient document,
21   Your Honor. More than 30 years old. Under the

22   Court's ruling, this comes into evidence.

23   ██████: Scientific literature isn't

24   automatically admissible under the ancient

191   ROUGH DRAFT TRANSCRIPT

1   document rule, Your Honor. We've talked about

2   this previously. It has to satisfy the learned

3   treatise exception because the witness is not an

4   expert. He's unable to satisfy that exhibit.

5   THE COURT: Ladies and gentlemen, I'm sorry

6   to have to give you a break already.

The simple fact that the witness has made a screenshot of the original historical source is insufficient for bringing it in as evidence. The fact that the chief medical officer is aware of it does not confirm for the court's satisfaction that it is admissible.

In short, in sprawling litigation where the past matters, it is no simple thing to bring evidence before a jury in the absence of a historian. The fact that a contemporary witness knows some history does not mean the court can allow the jury to see sources. Without a historian to identify evidence as evidence, it is possible that a jury may never even have an opportunity to encounter the past. In a situation like this, if the historian does not expertly witness history, the jury may have no opportunity to consider whether the past reflects reasonably foreseeable harms.

Conclusion

Expert testimony is sworn. While it is easy to be cynical about the limits of such avowals, I believe experts are either prepared or not. In the case of defendants, it may well be sufficient for them to hire an expert and prevent them from relying on a large number of materials. While such an elision appears suspicious, the conduct especially matters if the expert methodologically should consult a range of records.

For historians, then, matters are clear. Historians have to consult as much as is reasonably possible for them to consult and help clarify its contents. If they can bring evidence in through their testimony, then they can help ensure that the jury has an opportunity to hear evidence that might otherwise never become available to them.

Nonetheless, the court will place clear limitations on historical testimony. The court's exclusion of journalism, for instance, led in turn to an inability to bring into the jury's deliberations the fact that four of Matthew Gee's teammates were deceased. Equally, the court's determination that lawsuits, insurance records, and settlements are prejudicial to its proceedings can lead to an inability to rebut experts with facts that might unsettle the ease with which a jury accepts their reasonability.

Ultimately the courts create conditions that follow the law. In this sense the historian's desire for nuance is small in terms of the difficulty the jury faces. For the jury the matter boils down to a simple question: Is the defendant liable? It is inevitable that experts would disagree on that question, but in the absence of an expert witness for history, it is conceivable that a jury will never have an opportunity to contemplate whatever truth they determine the past plays in their answer.

NOTES

I thank Sharrona Pearl, Kelly O'Donnell, and the editors for constructive feedback on earlier versions of this chapter.

1. Amanda Christovich, "Where the NCAA's Wrongful Death Trial Stands after Two Weeks," Front Office Sports, November 11, 2022, https://frontofficesports.com/plaintiff-gee-trial/.

2. "NHL, Retired Players Reach $19M Concussions Settlement," *USA Today*, November 12, 2018, https://www.usatoday.com/story/sports/nhl/2018/11/12/tentative-settlement-reached-in-nhl-concussion-lawsuit/38484627/.

3. It is important to note that I am being compensated for my time. I disclose that I am retained in cases by plaintiffs in concussion litigation pending against sporting organizations globally.

4. Robert N. Proctor, "Should Medical Historians Be Working for the Tobacco Industry?," *Lancet* 363, no. 9416 (2004): 1174–1175.

5. Richard J. Evans, *In Defence of History* (London: Granta Books, 2001).

6. I find this argument in David J. Rothman, "Serving Clio and Client: The Historian as Expert Witness," *Bulletin of the History of Medicine* 77, no. 1 (2003): 25–44.

7. Constance Penley, "Collision in a Courtroom," in *Images, Ethics, Technology*, ed. Sharrona Pearl (London: Routledge, 2016), 58–72.

8. "Gee V. NCAA: Trial 10/21/22–11/22/22," CVN, accessed May 18, 2023, https://cvn.com/proceedings/gee-v-ncaa-trial-2022-10-10. The whole trial may be purchased on DVD.

9. "Athletes in Pregnancy Advised to Weigh Risk," *NCAA News*, February 27, 1985, 3.

10. Gee v. NCAA, Case No. 20STCV43627, Reporter's Transcript of Proceedings, Trial Day 19, Thursday, November 3, 2022, 154–155.

11. Gee v. NCAA, Reporter's Transcript of Proceedings, 154–156.

12. Gee v. NCAA, Reporter's Transcript of Proceedings, 155.

13. Michael Rosenberg, "In 1989, USC Had a Depth Chart of a Dozen Linebackers. Five Have Died, Each before Age 50," *Sports Illustrated*, October 6, 2020, https://www.si.com/college/2020/10/07/usc-and-its -dying-linebackers.

14. "Focus on Traumatic Brain Injury Research," National Institute of Neurological Disorders and Stroke, last reviewed July 19, 2024, https:// www.ninds.nih.gov/current-research/focus-disorders/focus-traumatic -brain-injury-research.

15. Gee v. NCAA, Reporter's Transcript of Proceedings, Trial Day 12, Tuesday, October 25, 2022, 74.

16. Gee v. NCAA, Reporter's Transcript of Proceedings, Trial Day 28, Thursday, November 17, 2022, 120–121.

17. Chris Murphy, *Madness, Inc.: How College Sports Can Leave Athletes Broken and Abandoned* (2019), 13, https://www.murphy.senate.gov /download/madness-inc-3.

18. Jennifer Finnerty v. NCAA, Undated Rough Transcript, 2023, 183–184.

PART IV
Writing

CHAPTER SEVENTEEN

Names

ADRIA L. IMADA

The name Annie Kekoa appeared in print in 1917. The writer Charmian London described meeting the young and "very charming" Kekoa during her visit to the Kalaupapa leprosy settlement on the island of Molokai, Hawai'i.[1] Kekoa, after being diagnosed with the bacterial disease known as leprosy, had been exiled by the Hawai'i Board of Health to this isolated peninsula. She was living in a community of 700 patient-inmates when Charmian London and her husband, the novelist Jack London, toured the settlement in July 1907.

At the time, leprosy was a stigmatized illness with an unknown and mysterious transmission. Although Charmian London pronounced Kekoa "without blemish," leprosy could cause dramatic disfiguration of the face, feet, hands, and skin, as well as blindness.[2] It affected people of all nationalities and ethnicities, but leprosy became associated with Native and nonwhite communities during nineteenth-century colonial expansion. Those who contracted leprosy were blamed for unsightly disabilities, disorderly behavior, communal households, and illicit sexuality. The Kalaupapa, Molokai, settlement in Hawai'i had become world famous for imprisoning people with leprosy. A cure for leprosy would not be discovered until the 1940s.

Annie Kekoa was the only person with leprosy identified by name in Charmian London's travelogue *Our Hawaii*. London reported a host of biographical details about Kekoa. Some inaccurate or unverified details notwithstanding, curious readers would have learned of Kekoa's hometown (Hilo, Hawai'i), her occupation (telephone operator), and her father's occupation ("native minister") while she was still living.[3] In an unusual reversal of Kekoa's lifetime medical sentence, the board of health released her back to her family around 1910. Due to the publication of her name in London's account, however, Kekoa would have been potentially associated with the criminalized condition of leprosy until her death in 1932.

It is unclear whether Kekoa gave London permission to report her personal information, but consent and privacy would not have been a consideration at the time. As she was a prisoner, Kekoa's medical information, photographs, and biographical data would have been available to board of health officials and distributed at their discretion. This is likely how Kekoa's clinical photograph from her 1905 examination made its way into Jack London's personal album, now held by the Huntington Library.

Seeing Kekoa's name in Charmian London's account was unsettling to me. On the one hand, Kekoa's name signified her relationship to a larger community beyond that of the prison. But I felt unsure whether I should pursue these connections. Could I write about Kekoa and thousands of other people whose lives were interrupted by medical sentences?

Every man, woman, and child determined to have leprosy in Hawai'i was subject to lifelong imprisonment: they became civilly dead, losing freedom of movement and access to their loved ones. The vast majority of the 8,000 leprosy patients sentenced to the Molokai settlement between 1866 and 1969 were Kānaka 'Ōiwi (Native Hawaiian) like Kekoa. Smaller numbers of immigrants from China, Japan, Korea, Portugal, and the Philippines were exiled, as well as their descendants.

I handled each person's examination record—approximately 1,400 extant files—over days and weeks that ended up stretching into years. I had privileged access to Kekoa's information due to the passage of time. Her file was unrestricted because the Hawai'i state statutes governing confidentiality had expired for her record.[4] Nor did HIPAA (Health Insurance Portability and Accountability Act) regulations apply, for it was more than fifty years after her death in 1932. Kekoa did not enjoy rights of privacy or consent during her lifetime. But should I use her name and those of other now-deceased inmates in my writing? If so, in what contexts and with what limitations? What injuries and harm might these disclosures cause? Should I rely on pseudonyms instead?[5]

Unlike Jack and Charmian London, who sojourned briefly in the islands, I am a settler born and raised in Hawai'i. I carry ongoing values and obligations to this home that fed and nurtured me: to care for the people I write about and their ancestors, collateral kin, and future relatives. How, then, to proceed? Guided by such connections, I arrived at a series of imperfect decisions to include some given names and familial information in my recent book, *An Archive of Skin, an Archive of Kin.*[6]

***

The board of health files at the Hawai'i State Archives are a paper-intensive set of holdings spanning the board's long existence. This public health agency began in 1851 as an entity of the Hawaiian Kingdom. Approximately thirteen years (1905–1917) of health correspondence managed to be archived. Within these files, I recognized names and relationships in letters written by family members of patient-inmates, including people within my larger community circles.

One folder contained letters written in Hawaiian by a man named Edward Kekoa to the then-president of the board of health. Edward was Annie Kekoa's father as well as a minister of a Protestant church in Hilo, Hawai'i. In his letters, it becomes readily apparent that Annie (written in Hawaiian as "Ane") was far more than Case 545 to her friends and family. Edward contested Ane's detention after her 1905 arrest by health officers. Writing in Hawaiian, Edward demanded of the president of the Board of Health, "E hookuu a hoihoi mai i kuu kaika-mahine ia Miss Ane N. Kekoa" [Release and return my beloved daughter Miss Ane N. Kekoa].[7] He expressed his love for his daughter and petitioned for her return from Kalihi Hospital, the leprosy detention facility in Honolulu.

Throughout the decades of Hawai'i's harsh leprosy incarceration system, such letters, petitions, and hospital visitation permits constitute an inextinguishable part of family reclamation and repair. Family members refused to relinquish loved ones to the colonial government—even if they were separated by laws, oceans, fences, and breath. People on the outside and exiled people living within the settlement alike placed the names of treasured kin in a printed public record: namely, in Hawaiian-language newspapers.[8] That these names appeared frequently across these newspapers in the early twentieth century suggests that it is possible to disclose their names while imparting the texture of their lived experiences, desires, and subjectivities.

In printed kanikau (mourning chants) and poetic remembrances, Hawaiian people entreated a broad community of readers not to forget their spouses, friends, and cousins who had been removed to the Molokai settlement. Their specific biographies, hometowns, names, ancestors, and kin often were included in these remembrances. Among many instances, a woman signing her name as Mrs. H. P. Paniani submitted a moving tribute in 1912 upon the death of her friend, a Mrs. Kalamau who was originally from Pahala, Kaʻū, Hawaiʻi. Both women were patients residing at Bishop Home, the women's dormitory in the Kalaupapa, Molokai, settlement.[9] Other exiled people sent updates to newspapers with the names and activities of fellow compatriots, telling the outside world of the pleasures and hardships in their settlement community in the early 1900s. They freely reported biographical details, ages, names, and homelands. They did not want to be forgotten.

Indeed, people struggled during medical incarceration, which reduced them to a disease, disability, or diagnostic condition. It almost made it impossible for a person to be seen as human. A Hawaiian man exiled around 1920 succinctly expressed his transformation from a schoolboy into a loathsome Other at the Kalihi detention hospital: "So I became a leper."[10]

At a similar, but later, institution for leprosy patients—the US national leprosarium in Carville, Louisiana, founded in 1921—the shame and disgust associated with leprosy were so great that people passing through its doors were encouraged to relinquish their birth names and adopt "Carville names" to protect their families. Internalizing the stigma, many retained their Carville aliases even after mandatory quarantine in Louisiana was lifted in the late 1950s. The writer known as Betty Martin published two popular memoirs about her residence at Carville under an alias. Martin, a white woman from New Orleans, refused to reveal her first and last name seventy years after her diagnosis.[11]

At Kalihi Hospital and Kalaupapa settlement, people did not change their names. Yet the past practices of exiled people like Mrs. Paniani are important clues and guideposts. They suggest that recording the names and attachments of patient-inmates can serve as a partial challenge to the violence and dehumanization that ensnared them. Retaining these names helps to reinforce the personhood stripped away by disease and colonial medical institutions. Doing so can convey a broader sense of kinship, birthplace, and relationships before and after a person's entry into biomedical institutions. While perhaps providing only an incomplete portrait, these naming practices convey a person's fuller lifeworld beyond that of a medical record or clinical setting.

Might aliases be another solution? The use of pseudonyms is not ideal in a Hawaiian context. Hawaiian names are neither easily endowed nor excised, for it is believed that inoa (names) hold deep power that can shape a person's fate and misfortune. Names also reveal how people thought of themselves, sustained themselves creatively, and rebuilt social environments. The Lapilio family in the early 1900s forged new relationships through names. Ponepake Lapilio, a young man exiled without his family in 1889, was cared for by an older man named Bernard Palikapu upon his arrival in Molokai. When Lapilio's wife bore a second child in 1905 at Kalaupapa settlement, the couple named the baby Bernardo Palikapu after their older friend, thus cementing a relationship between generations.

On the other hand, should all aspects of people's lives be accessible to modern researchers? I have heard a prominent researcher advocate letting go of privacy in histories of medicine; in other words, the dead have no claims to privacy. This makes me deeply uncomfortable. Those who are no longer living are far more than names and "data": they are people who continue to be remembered and claimed by living relations.

While I usually hew to given names in my writing, I also exercise restraint. I refer to this practice as an *ethics of restraint*—a deliberate attempt to refrain from overexposing and spectacularizing vulnerable people. Just because one *can* does not mean one should. For if names tell important stories, they can also hold secrets. Some of the people I encountered through conversations and paper records experienced severed relationships, estrangement, and shame. I may have chosen them as my subjects, but they did not choose me as their storyteller. Acknowledging the asymmetry of these relationships forces a pause. As the narrator, I hold more power of access, choice, and voice. Therefore, I may allude to social trauma while avoiding further injury. When left to interpret oblique silences or the avoidance of elderly people, I err on the side of restraint and omission as well.

A careful use of names, accompanied by transparent explanations, then, ideally is aligned with justice-oriented language. This language centers the desires of those most affected. Indigenous survivors continue to live with the compounded consequences of public health racism, settler-colonial violence, and institutionalized removal from families.[12] Generations later, Kānaka ʻŌiwi (Native Hawaiians) and Indigenous people have poorer health outcomes and lower life expectancy compared with non-Natives. Disability justice and decolonization politics both urge attention to the language people prefer to be used to refer to them and how this may shift over time.

What were people's preferred self-ascriptions? In my writing, I generally eschew the label *leper*, which many in twentieth-century survivor communities found degrading. By the 1960s and 1970s, survivors advocated for the shift to the term *Hansen's disease* from *leprosy*. Yet even within this serious context of naming, one exiled man joked and called his condition "leperoses" or the "handsome disease." And what of the flawed term *patient*? I chose to rely on *patient* although it may retain a patronizing, medicalized tone to some ears. For among Kalaupapa settlement survivors and caregivers, *patient* remains the colloquial, yet respectful, term for elders who survived harsh isolation and incarceration.

Despite these speculative boundaries and guardrails, no choice is perfect. I often sit with unease, wondering whether I have made the best decisions or trespassed in ways not yet apparent. These remain vexing, but critical, issues for historians of medicine and disability; indeed, for anyone who attempts to work ethically on and with subordinated communities.

Colleagues have arrived at other meaningful decisions and protocols in their particular contexts of engagement. Susan Burch only included a person's name in her research on Native American women incarcerated at South Dakota's Canton Asylum if she received direct permission from lineal relatives. Alice Wexler started off using pseudonyms in a history of Huntington's disease, a fatal genetic disease, but later shifted to the real names of deceased people to lessen stigma. Indigenous and non-Indigenous activists, scholars, and archivists researching thousands of children incarcerated at federal Indian boarding schools and residential schools in the United States have explained their paths and consultation process with tribal leaders. Two researchers for an internet database of students at the Carlisle Indian School, for instance, held back some of the children's names from public view but made them accessible to particular tribal nations.[13] Not everyone has the right or responsibility to access these names.

The words and names we choose do matter to communities of survivors, whether they are relations or those who become affiliated through shared conditions of illness and disability. We ought to care more whether our words cause injury or pain. But I also urge contemplation of the *joyous* possibilities enacted by names. Names are gifts that can deliver joy, relief, comfort, recognition, and connection. I am reminded of these possibilities when listening to the kin of former patients. Recently a woman told me how she took comfort reading about her family in my book. She had learned for the first time how her forebears had relocated from another island to remain close to their children taken to the leprosy detention hospital in Honolulu. These strenuous efforts of her great-great-grandparents to hold on to their children had not been part of her history. But more than a hundred years later, their names on the page continue to endow love and care.

NOTES

Note on language: I do not italicize words in ʻōlelo Hawaiʻi (the Hawaiian language) to avoid marking an Indigenous language as foreign.

1. Charmian Kittredge London, *Our Hawaii* (New York: Macmillan, 1917), 128.

2. London, *Our Hawaii*, 128.

3. London, *Our Hawaii*, 128.

4. Hawaiʻi Revised Statutes § 92F-14 restricts access to confidential personal information in historical government records. All restrictions are removed eighty years after the creation of the record (Hawaiʻi Revised Statutes § 94-7).

5. As Michaela Clark discusses in this volume, similar considerations and choices accompany the use of clinical photographs in historical narratives. On the contextualization of photographs and historical materials while teaching students, see Beatriz Pichel's essay this volume.

6. Adria L. Imada, *An Archive of Skin, an Archive of Kin: Disability and Life-Making during Medical Incarceration* (Berkeley: University of California Press, 2022).

7. Edward Kekoa, letter to L.E. Pinkham, October 28, 1905, Folder: Kalihi Hospital, 1905–1909, Box 335-13, Correspondence of the Board of Health, 1905–1917, Series 335, Hawaiʻi State Archives. The quotation is from Kekoa's letter written in Hawaiian and is followed by my translation into English.

8. Despite the suppression of the Hawaiian language in education and public life after the 1893 US overthrow and 1898 US annexation, Hawaiian newspapers were still prevalent throughout the territorial period. There were ten in print as of 1909. Esther K. Mookini, *The Hawaiian Newspapers* (Honolulu: Topgallant, 1974).

9. "Hoomanao Ana i Ka Mea i Hala," *Ka Nupepa Kuokoa*, October 18, 1912.

10. Ted Gugelyk and Milton Bloombaum, *Maʻi Hoʻokaʻawale: The Separating Sickness* (Honolulu: Social Science Research Institute, University of Hawaii, 1979), 80. The word *leper* is generally avoided by patients today in favor of *Hansen's disease*. I do not use the former except when quoting directly from survivors or from historical documents.

11. On changing one's first and last names upon entry to Carville national leprosarium, see Stanley Stein [Sidney Maurice Levyson], *Alone No Longer: The Story of a Man Who Refused to Be One of the Living Dead!* (New York: Funk and Wagnalls, 1963). Betty Martin's two memoirs are *Miracle at Carville* (Garden City, NY: Doubleday, 1950) and *No One Must Ever Know* (Garden City, NY: Doubleday, 1959).

12. Mary Jane Logan McCallum, "Starvation, Experimentation, Segregation, and Trauma: Words for Reading Indigenous Health History," *Canadian Historical Review* 98, no. 1 (2017): 96–113.

13. Susan Burch, *Committed: Remembering Native Kinship in and beyond Institutions* (Chapel Hill: University of North Carolina Press, 2021); Alice Wexler, *The Woman Who Walked into the Sea: Huntington's and the Making of a Genetic Disease* (New Haven, CT: Yale University Press, 2008); Barbara Landis, "The Names," in *Carlisle Indian Industrial School: Indigenous Histories, Memories, and Reclamations*, ed. Jacqueline Fear-Segal and Susan D. Rose (Lincoln: University of Nebraska Press, 2016), 89–90.

CHAPTER EIGHTEEN

Diagnoses and Language

CLAIRE D. CLARK AND AMY C. SULLIVAN

Diagnoses are freighted with power—the power to treat and hopefully to heal, but also the power to classify the patient and consequently to influence their sense of identity. A diagnosis can provide the relief of a new explanatory framework that makes order out of the chaos of the illness experience, but it can also alienate the patient from both the doctor and the self. As historians of medicine, we make choices about how to refer to our historical subjects' relationships to their health. The decision to reject the clinical gaze by using the language more commonly used to describe medical conditions in everyday speech is not necessarily a liberating alternative, as this language can also serve the needs of various other stakeholders in the patient's social position and perpetuate stigma.

Consider the word *addict*. Historian and clinician William L. White discusses the development of labels for people who use alcohol and other drugs since the late 1700s in the United States. White argues that the "confusion and conflict surrounding this evolving language" used to label people with problematic relationships with substances stem from the fact that the words used to describe them often serve several purposes simultaneously: personal utility, social and political utility, professional utility, clinical utility, and economic utility.[1] We believe that the language we adopt in our historical scholarship should, first and foremost, have personal utility to our patient-protagonists.

As an exercise in language and utility, this essay focuses on the use of the word *addict* in historical writing about people who use drugs. The term *addict* has persisted in both humanities scholarship and the broader culture even as other similarly stigmatizing words have fallen out of favor. The shift to person-first language in disability rights organizing in the 1970s and 1980s, and later in medical and therapeutic settings, was an important step toward humanizing patients in increasingly sterile, corporate medical systems—it focused attention on people and not just their afflictions, especially among historically marginalized and stigmatized populations. But for people with substance use disorders, person-first language has not been incorporated into public, familial, and medical settings as quickly (if you can call decades later *quickly*). One reason may be related to the ubiquity of Alcoholics Anonymous practices, where participants have historically referred to *themselves* as "addicts" and "alcoholics," in part to signify a nomenclature of shared lived experiences and a reminder of the diligence required to maintain sobriety. The organization's commitment to anonymity may have also played an unintended role in the perpetuation of the use of these two words in our broader culture. To outsiders, the social invisibility of people in recovery has kept intact the ubiquitous use of words that we now know perpetuate stigma.[2] By dropping the use of *addict* in history writing, teaching, and scholarship, historians can play an important role in centering and humanizing our subjects, living or dead. We believe this practice is part of our responsibility as humanists to improving the world we currently inhabit.

In the midst of an overdose crisis that has claimed more than a million lives since 1999, we write as both historians and people with personal connections to addiction—and as people who believe that our positionality informs our histories.[3] We call for abandoning the use of the term *addict* in favor of adopting person-first descriptions instead.[4] The word *addict* was initially used in the professional literature at the beginning of the twentieth century to refer to people who used a range of substances, while *addiction* itself was not empirically defined until human subjects research was conducted at a federal hospital and prison that treated people who used narcotics (e.g., heroin) beginning in the 1940s. By the mid-twentieth century, though, the word *addict* became associated with people who used illicit drugs, while the word *addiction* expanded to encompass the problematic use of alcohol and tobacco, as well as an ever-expanding range of other compulsive behaviors.[5] In the pages that follow, we reflect on our own historical work and reject the argument that the anachronistic and pejorative term *addict* has any continued historical utility.

Our Work

Claire Clark: My perspective as a person in recovery influenced my choice of historical subject and my use of language even as I tried to position myself as a neutral observer of the often colorful history of alcohol, drugs, and addiction in twentieth-century American medicine. In a prominent lecture delivered to the American Association for the History of Medicine, Susan Reverby urges her colleagues to write "toward a passionate history"—history that strives for ever-elusive "balance" and careful contextualization while at the same time acknowledging that a writer's particular passions will inevitably shape their descriptions of historical actors.[6] She describes the challenging process of writing about "two great(ly infamous) doctors," John C. Cutler and Alan Berkman. Both men were complex: "Cutler, admonished for his role in the infamous sexually transmitted diseases studies in Tuskegee and Guatemala, was also a well-respected researcher and teacher. Berkman, renowned for his success in global HIV/AIDS activism, was also only the second physician in U.S. history to be charged with accessory to murder after the fact and who served seven hard years for bombings and robbery."[7] In writing about these men, Reverby worried that her aversion to Cutler's abhorrent actions would impair her ability to contextualize how he operated within the racist and colonialist mores of the technocratic liberalism of his time, while her sympathy for the leftist Berkman, whom she personally knew, might temper her condemnation of his youthful commitment to revolutionary violence. In either case, Reverby argues, writing an objective and purely dispassionate history of these doctors was both an impossibility and an ideal that historians should be willing to relinquish. An awareness of subjectivity and positionality is integral to any attempt at historical balance.

I have thought about Reverby's advice in both teaching and researching the history of the "Lexington Narcotic Farm," an infamous icon in the history of addiction treatment in the United States. One of two such hospitals in the country, this federal hospital and prison was, like the Tuskegee Syphilis Study, run by the US Public Health Service from the 1930s until the 1970s. It was located in Lexington, Kentucky, and had ties to the University of Kentucky, where I now teach; I even had an opportunity to tour the facility as the director of a National Endowment for the Humanities Institute on Addiction in American History in 2018.

In the mid-twentieth century, Lexington—or Narco, as it was sometimes called—served as one of the few institutions in the country designated for the treatment of people who used narcotics (then defined as opiates, cocaine, and marijuana). In the words of one historical survey of the institution, it also housed the "only laboratory in the world that had access to a captive population of highly experienced and knowledgeable drug addicts, many of whom were more than eager to participate in experiments involving drugs of any kind."[8] With few exceptions, historians of medicine who have written about Lexington have centered the physician-scientists who worked there rather than the so-called addicts who were supervised by them.[9] They have striven to articulate the logic of prisoner research that would be found to be inherently coercive and unethical by the 1970s.[10] Only recently has the history of experimentation in Lexington been told empathetically from the perspective of someone who is a person in recovery as well as a clinician and who is able to fully articulate the silences and violences of the infamous research.[11] Carl Erik Fisher writes,

> In a series of experiments that would make today's ethicists blanch, researchers gave the patients opioids until they developed tolerance, then plunged them into withdrawal by stopping the drugs (or later by giving them opioid-blocking medications). As the patients groaned, the researchers stood by with clipboards, dutifully tracking all the objective signs of drug use and withdrawal to the minute: heart rate, yawning, sweating, muscle twitches, goosebumps, diarrhea, even spontaneous orgasms. To measure pain and suffering, scientists put hot metal on the subjects' skin or administered electric shocks to their teeth, asking them to quantify the results.[12]

The research conducted at Lexington in the mid-twentieth century exploited the knowledge and expertise of prison inmates who use drugs and defined what it meant to be an "addict." In teaching graduate students and postdoctoral trainees who are focused on careers in substance use research, I have introduced community-driven research as evidence of a promising historical shift from doing research *on* people who use drugs to doing research in partnership *with* people who use drugs.[13] Such partnership necessarily involves researchers and participants co-creating knowledge about the lived experience of substance use using mutually agreed-on terms. In this more recent teaching of trainees in substance use research, I have also begun positioning myself as both a historian of medicine and a privileged person in long-term recovery from problems with food and alcohol. I now share that I am growing from privately and anonymously identifying as a recovering addict to professionally identifying as a harm reductionist—a grateful recovering human who has long been coping with life, in all its messy complexity, to the best of my ability.

My previous writing focused on what I called the recovery revolution, a movement in which self-described "ex-addicts" propagated a controversial, peer-led, abstinence-based treatment paradigm that was imagined in contrast to, but ultimately became entwined with, the technocratic liberalism of carceral-therapeutic institutions such as Lexington.[14] I opened the preface of my book, published in 2017, by placing derogatory language used to describe people who use drugs in quotes, yet in the book's body I occasionally dropped the quotation marks and used the word *addict* in my historical descriptions. For example, I wrote, "As a technocratic project, the narcotics farms' failed treatment experiments reaped distressing results. Officials labeled drug users 'addicts' and corralled them into two central locations: once in those locations, the addicts internalized the labels."[15] In my attempt to craft a balanced yet propulsive historical narrative from oral histories and archival documents, I also internalized the labels; I would never have used the word *addict* in my public health writing at the time. I am committed to finding more humane alternatives in my future historical work.

***

Amy Sullivan: Before I began researching the current opioid epidemic, I was a regular attendee of a Twelve Step–focused family support group for loved ones of people who used drugs harder than alcohol. In these spaces, members often referred to their loved one as "my addict," but this moniker never sat well with me—it felt simultaneously possessive and stigmatizing. Years earlier, I had experienced the same bristling feeling in the ICU when my loved one was quickly dismissed as "an OD" to a group of medical residents. It seemed that in the very places where we most need humanity, this kind of naming dehumanized all of us.

What felt disconcerting and stigmatizing to me in my personal life was soon reinforced in my scholarly research by the more than fifty individuals I interviewed whose lives and work intersected with the consequences of the opioid epidemic. As I listened, I became sensitized to the range and meaning of words and expressions that individuals used to describe themselves and their experiences. As William White explains, "The language of addiction is a coded language. . . . The rhetoric chosen to define and discuss alcohol and other drug addiction itself defines addicted people in certain ways and rationalizes particular types of interventions into their lives."[16] To this I ask, what is our responsibility as health and medical historians to the people we write about? How do we give them agency in historical scholarship, regardless of what they were called in their time? If one of our goals is to do less harm, I argue that humanities scholars have an ethical responsibility to not just show what happened in the past but also use language that creates space and distance, rather than words that reduce their experiences to something that remains easy to stigmatize in the split second it takes to read it. Although it has become more common for historians to provide a "note about language" apology for using these historical terms (whether they were claimed by the subjects or not), the language of the time that they employ in their writing has the effect of reinforcing its value and use in the present. Even long-dead historical subjects should be allowed more neutral language. We can use our scholarly and professional privilege to model person-first histories in as meaningful a way as health care and human service providers do now with their growing adoption of person-first language.

Two examples from physicians in my oral history collection demonstrate how easy it can be to drop this language. In his addiction medicine practice, Mark Willenbring distributed a handout to patients entitled "Vocabulary Means a Lot." Instead of *addict,* he suggested *person with a use disorder*; instead of *relapse,* consider *recurrence* or *setback.* He preferred the phrase *withdrawal management* over *detox.* "They're not *toxic!*" he exclaimed to me.[17] Another physician who directs the addiction medicine clinic at a large county hospital changed his language completely in the span of two years—from the time of my interview with him to the completion of my manuscript. At some point during my research, I discovered that he used the word *addict* about twenty times in the interview.[18] Two years later, when I asked him to review quotes and content of his that would be in the book, he wrote back to say he had cringed at reading his use of the word *addict* then. "I don't doubt that I did, but it is now considered poor form."[19]

As writers and historians, we must also be attentive to a duality: the way certain language is and has been used to stigmatize and stereotype people as well as moments when those very same historical subjects claim or reclaim a word for themselves as an essential part of their identity. When one narrator regularly referred to himself as "an old school drug addict," I asked him if he had any problem with others using the word *addict* when referring to him. He said, "I believe if you have to do things every day to make sure you don't use, then I think you are an addict. The people who have a problem with language issues aren't serious drug addicts anyway. If you have had a lot of time using and trying to get clean, you aren't worried about someone calling you an addict. That is 'little shit,' and it doesn't bother me."[20]

Whether it is a living person across the table or a life discovered in an archive, the distance between the historical subject and the writer really makes no difference when it comes to how historians write about subjects in our own time, for our own audiences. Using agentive, respectful language allows readers the opportunity to see historical subjects as much on their own terms as possible. Historical "distancing" through language—for example, not using *addict*—is one way to provide space for our subjects to be the complex historical actors they are or were, which, whenever possible, includes allowing them to use descriptions of themselves that feel true and honest.

When I first began collecting oral histories from people in Minnesota who had been affected by the opioid epidemic, I used the word *addict* myself a few times. But four years and fifty interviews later, I decided not to use the word in my life, my classrooms, or my writing. I now also ask my students to do the same. This slight shift in language usage has changed the tenor of our discussions. Now students notice when academics and journalists use the word *addict*. They feel the difference, too, and comment on the empathy created by this simple act. Once they have stopped using the word, they are doubly horrified by histories of medical injustices as well as by the way historians write with clinical distance about shocking abuse and maltreatment. This experience in pedagogy solidified my commitment to changing language and labels. The language we use has the ability to cultivate empathy for people in the past and the present—students, scholars, and policymakers alike. In fact, the students who take history of medicine courses are often the very people who will be treating patients in the future—our futures! Could person-first language help bridge the gulf between our discipline's focus on the past and our present-focused students? If person-first wording feels clunky, which sometimes it does, consider using any word that is neutral and not bogged down in stigma and stereotypes. One example comes from a former fentanyl user who prefers to say "when I got healthy," in lieu of *recovery*, because using the word *recovery* often connotes Twelve Step programs, which she does not participate in. Using different words to describe stale, overused ones provides a fresh look at our subjects and may also have the same positive effect on our writing.

In Conclusion

Language and labels are ever evolving, and it is possible, perhaps even likely, that person-first language in the history of medicine will eventually be replaced by a more sensitive future alternative. As our examples demonstrate, continuing to use the word *addict* in historical writing does harm to both our immediate readers and the wider public. Using *addict* in writing directed toward small audiences of highly specialized historians of medicine is not inclusive; it assumes that people who struggle with substances are not among our readers. The word *addict* additionally does harm to the wider public by perpetuating stigma. When we do less harm in our writing by rejecting the reductionism of diagnostic classifications or pejorative common language, we are also respecting our readers. Historians seeking to center people struggling with their relationships to substances should find other ways to describe their actors and narrators. In fact, all health historians should consider engaging with the communities under discussion about language and remain attuned to checking whether those labels overgeneralize or stigmatize our subjects.

NOTES

1. William L. White, "The Lessons of Language: Historical Perspectives on the Rhetoric of Addiction," in *Altering American Consciousness: The History of Alcohol and Drug Use in the United States, 1800–2000*, ed. Sarah W. Tracy and Caroline Jean Acker (Amherst: University of Massachusetts Press, 2004), 33.

2. Amy F. Crocker and Susan N. Smith, "Person-First Language: Are We Practicing What We Preach?," *Journal of Multidisciplinary Healthcare* 12 (February 2019): 125–129.

3. See Barron Lerner's essay in this volume.

4. "Words Matter: Preferred Language for Talking about Drug Addiction," National Institute on Drug Abuse, June 23, 2021, https://nida .nih.gov/research-topics/addiction-science/words-matter-preferred -language-talking-about-addiction; "Preferred Terms for Select Population Groups and Communities," Centers for Disease Control and Prevention, November 3, 2022, https://www.cdc.gov/healthcommunication/Preferred _Terms.html.

5. White, "Lessons of Language."

6. Susan M. Reverby, "Enemy of the People/Enemy of the State: Two Great(ly Infamous) Doctors, Passions, and the Judgment of History," *Bulletin of the History of Medicine* 88, no. 3 (2014): 403–430.

7. Reverby, 403.

8. Nancy D. Campbell, J. P. Olsen, and Luke Walden, *The Narcotic Farm: The Rise and Fall of America's First Prison for Drug Addicts* (Lexington: University of Kentucky Press, 2021), 162.

9. See, for example, David Musto, *The American Disease: The Origins of Narcotic Control* (Oxford: Oxford University Press, 1999); Caroline Acker, *Creating the American Junkie: Addiction Research in the Classic Era of Narcotic Control* (Baltimore: Johns Hopkins University Press, 2003); Nancy Campbell, *Discovering Addiction: The Science and Politics of Drug Abuse Research* (Ann Arbor: University of Michigan Press, 2007). For oral histories that do capture residents' perspectives of their time at the Narcotic Farm, see David Courtwright, Herman Joseph, and Don Des Jarlais, *Addicts Who Survived: An Oral History of Narcotic Use in America before 1965* (Knoxville: University of Tennessee Press, 2012).

10. Nancy Campbell and Laura Stark, "Making Up Vulnerable People: Human Subjects and the Subjective Experience of Medical Experiment," *Social History of Medicine* 28, no. 4 (2015): 825–848.

11. See Ayah Nuriddin's essay in this volume.

12. Carl Erik Fisher, *The Urge: Our History of Addiction* (New York: Penguin, 2022), 189–190. The "Narco formulation" was the classificatory scheme the researchers used to explain the physiological basis of opioid dependence and withdrawal.

13. Michael J. Montoya and Erin E. Kent, "Dialogical Action: Moving from Community-Based to Community-Driven Participatory Research," *Qualitative Health Research* 21, no. 7 (2011): 1000–1011; Caty Simon et al., "We Are the Researched, the Researchers, the Discounted: The Experiences of Drug User Activists as Researchers," *International Journal of Drug Policy* 98 (2021): 103364.

14. Claire D. Clark, *The Recovery Revolution: The Battle over Addiction Treatment in the United States* (New York: Columbia University Press, 2017).

15. Clark, 9.

16. William L. White, *Slaying the Dragon: The History of Addiction Treatment and Recovery in America,* 2nd ed. (Bloomington: Chestnut Health Systems, 1998), xxi.

17. Mark Willenbring, transcript of interview by Amy C. Sullivan, May 17, 2017.

18. Charles Reznikoff, transcript of interview by Amy C. Sullivan, January 17, 2017.

19. Email correspondence with Amy C. Sullivan, August 20, 2019.

20. Amy C. Sullivan, *Opioid Reckoning: Love, Loss, and Redemption in the Rehab State* (Minneapolis: University of Minnesota Press, 2022), 218.

CHAPTER NINETEEN

Writing about Images

MICHAELA CLARK

Clinical photographs are simply defined as images made with a camera that depict living human subjects suffering conspicuous symptoms of disease.[1] Today, these photographic renderings of patients and pathology operate as personal data. They are therefore regulated like any other part of a person's medical record. While ethical guidelines depend on geographical context, the principles that consistently inform the making and use of clinical photographs revolve around the protection of privacy. But regulations of this kind only mediate images of *living* individuals. Historical material (depicting those long dead) falls outside this medico-legal remit. Long harnessed for medical recordkeeping, as teaching aids, and as illustrations for research and publication, many of these photographs were produced before the implementation of informed consent.[2] Without relevant regulations or the possibility to establish consent in retrospect, it often falls to the historian to make an executive decision on whether it is ethical to reproduce photographs of patients past. This chapter prompts readers to reflect on their position and the choices they might make when deciding to use (or not use) material of this kind.

Looking at Clinical Photographs

To pose an obvious question, Why would historians wish to look at clinical photographs in the first place?

1. Historical Artifact

Although made with clinical ends in mind, photographs of patients offer unique primary source material. As products of institutional behavior and bureaucracy, medical records can operate as documents with "intrinsic or historical value" because they offer a window into administrative practices and conceptual frameworks of the past.[3] As visual sources, clinical photographs evidence how patients were looked at within medicine. They thus help to bring the medical gaze into focus while also offering traces of the social, cultural, and political influences on the practices of medicine at any given time.[4] Having fallen into disuse as medical records, clinical photographs can thus provide insight into a wider intellectual history.[5] This is what makes them so useful for the historian as well as for those teaching the history of medicine, as Beatriz Pichel shows in this volume.

2. Patient Experience

Medical records more broadly facilitate "patient testimony, protest, and indeed their own experiences of confinement."[6] They thus offer an opportunity for the historian to "retrieve" patients of the past. For, as images of people, clinical photographs hold details that reveal not only medical norms but individual human experience. This is because they frequently include information that exceeds the intentions of the photographer. Who patients were, how they were handled, and even (at times) how they may have felt can be revealed through posture, dress, and facial expressions.[7] Those depicted are thereby prevented from "falling into oblivion" or from being "erased from the historical record and memory."[8] Clinical photographs thus allow the historian to tell the history of medicine "from below" (from the perspective of the patient) and thereby challenge "official" histories of scientific professionals.[9]

3. Public Impact

Sharing patient experience through photographs may serve to destigmatize disease and disability.[10] This is because medical images of this kind are "objects par excellence in recounting disease stories."[11] As such, they can help create personal narratives of illness and the medical encounter because photographs render experience intelligible when carefully exhibited. As Mieneke te Hennepe advocates with regard to the medical museum, this space "can be a platform for debate to serve public interests and communities[,] . . . reshape public consciousness of medical practices then and now, [and] urg[e] us to rethink the position of (marginalised) patients . . . or any other 'victimised' community in society."[12] By making extraordinary bodies visible and comprehensible to a broader public, depictions of patients can have a "far-reaching social impact" because they highlight affected people as individuals with rich emotional lives rather than simply as "sick" or "diseased."[13]

4. Visibility

A core purpose of archives and museums is to do "memory work."[14] This means aiding individuals, institutions, and societies to grapple with the past. Included in such efforts is the writing of reparative histories, as articulated in this volume by Kylie Smith. In societies and communities that have experienced state oppression or suffered collective trauma, the most vulnerable may be otherwise invisible to the historical record and actively seek acknowledgment.[15] This is why many Holocaust survivors and their descendants advocate for the opening of medical records of patients affected by Nazi-era medicine. The motivation for such action is to ensure justice is served, to gain a sense of personal closure, and to facilitate remembrance.[16] Photographs may further offer legal evidence, with some advocates affected by the thalidomide disaster requesting the publication of their images.[17] Here visibility operates as proof of institutional injustice and medical malpractice. Making this material accessible ensures restitution while also offering a warning to industry: do wrong and you will be held to account.[18]

5. Accountability

Archives do not only preserve records; they "protect the rights and benefits of all citizens" by ensuring those in power are held "accountable for their actions."[19] Closure operates as a form of censorship that can allow institutional legacies to be shielded from critique. Indeed, removing access to historical scientific collections risks fostering "a conspiracy of silence" that forgoes the possibility of reparative action.[20] Thus, in terms of photographic records, even Susan Sontag—a longtime critic of photography—suggests that society has an obligation to forgo "innocence, . . . ignorance, or amnesia."[21] The ethical avenue would see society "haunted" by the photographically depicted events of the past. Historians thus hold a "civic responsibility and duty of care" to make sure that the disenfranchised are included in historical tales of medicine, health, and healing.[22]

6. Emotion

For Judith Butler, it is the *emotive* potential of the medium that allows photographs to prompt a deep recognition of human vulnerability and the value of human life.[23] They are affective images or "objects of feeling and relation," as they "allow us to feel things we don't quite understand, they make us dig, and even think, a little longer, a little deeper."[24] Clinical photographs are particularly forceful in this regard because they "bring us into uncanny proximity with the dead and the dying."[25] It is their "immediacy, graphicness, and affective aspects" that allow them to highlight the fragility of the human body, forcing viewers to confront their own physical frailty and, ultimately, their own mortality.[26] By offering a glimpse of corporeal collapse, medical images of this kind thus prompt a powerful means of photographic witnessing.

7. Reclaiming

Historical medical materials contain "information [that] is unique and very valuable to capture the daily life of our ancestors."[27] While produced toward medical ends, their meaning is malleable, in line with shifting contexts, custodians, and viewers. As demonstrated by Adria Imada in this volume, kinship ties to medical records can challenge the objectifying and dehumanizing nature of colonial medical practices.[28] Photographs in particular create a powerful sense of proximity to loved ones long dead. When transferred from the space of medicine to that of the family home, clinical photographs from World War I have prompted a deep sense of connection and care from living relatives.[29] Here, personal memory sees such institutional records "recovered" and "reclaimed" as familial images.[30] Similarly, in 2019, Harvard's slave daguerreotypes were claimed by the descendants of those depicted to function as "stolen family property."[31] Despite being made in 1850 as scientific evidence, these items were conceptually and emotionally transformed by intimate attachment from institutional evidence into personal tokens.

Not Looking at Clinical Photographs

With all of these reasons for historians to use clinical photographs, what is the cause for caution? Why should we *not* look at medical images of this kind?

1. Exposure

Clinical photographs are exposing in that they offer visual evidence of a patient's personal, and at times embarrassing, struggle with disease, and they often reveal unclothed intimate areas.[32] A key worry is therefore the use of these photographs beyond their original clinical remit (to which the patient may have agreed). In addition, these depictions reveal individual identity—the key legal issue when it comes to their circulation.[33] Identification may have negative consequences for familial or cultural descendants of those imaged, as living relatives can be harmed by having their heritable health information revealed.[34] Ethical concerns thus extend beyond death and even beyond the patient on display.

2. Vulnerability

Historically, clinical photographs have tended to depict individuals whose social or economic existence makes them susceptible to not only being photographed but having their medical likeness circulated beyond their reach. This is especially true for medical images showcasing exceptional or anomalous bodies. Those with "demonstrable physical differences" have frequently served as medical exemplars if not curiosities.[35] Codified as pathological and thus abnormal, they are presented as in need of "genetic reconstruction, surgical normalization, therapeutic elimination."[36] This tendency can also be seen in the medium of photography more generally. Often it is members of the laboring classes, women, and people of color who are singled out for visual documentation. While people who are ethnically white, are middle class, and otherwise conform to social as well as bodily norms may escape the institutional lens, the bodies and minds of "others" are vulnerable to being subject to a disciplinary gaze that involves the camera.[37]

3. Violence

Clinical photographs raise important questions about who is depicted and seen versus who gets to do the seeing. Even photographs taken for altruistic purposes can embody a form of symbolic violence. Those imaged are oppressed once by the social order and again by the camera, which visually fixes them as victims of their disenfranchisement. Such subjects are, to paraphrase Abigail Solomon-Godeau, therefore doubly subjugated in photographs that showcase their suffering.[38] Clinical photographs raise vital questions about reproducing harm due to the social, scientific, and representational circumstances of those on display.[39] For, in these images, patients are frozen in their circumstances and without the ability to extract themselves from this frame. In comparison, viewers have the privilege of being outside the experiences shown. Thus, as Suzannah Biernoff states, "the wish not to be represented—not to be exposed, or made public, or held up as an example, or pitied, or studied—is surely something to be mindful of."[40]

4. Objectification

Clinical photographs are produced within the medical terrain to capture decipherable markers of disease. By directing a viewer's gaze toward purely clinical ends, these images are structured to exclude extraclinical elements. In essence they attempt to re-create the medical gaze of the clinic in image form. For Ian Berle, the problematic nature of clinical photographs is that the very act of "taking" the photograph "suggests ownership of something collectable or memorable, . . . increas[ing] the objectification of patients and reinforc[ing] the clinician's perceived authority."[41] The vantage point granted by the camera is inherently aligned with the norms and needs of the photograph's origin and maker. By focusing attention solely on the signs of disease, the photograph ultimately makes patients into medically useful "things."

5. Voyeurism

As visual records, photographs raise the possibility of enjoyment in looking. Frequently, depictions of extraordinary bodies teeter between scientific specimen and spectacle.[42] There is something fascinating about them, and interest in this medical material may be framed as (at best) gratuitous or (at worst) morbid. As Laura Mulvey conceptualizes it, enjoyment in seeing may be made manifest as a form of "pleasure" by allowing viewers to revel in accessing what should be private without the threat of being looked at in return.[43] The result is a form of voyeuristic gratification that subjects those on display "to assumptions which have nothing to do with [their] individuality."[44] Naked, sick, and exposed, patients depicted in photographic form are transformed into objects of visual pleasure for the viewer. Turning even the most ordinary or horrific scene into something pleasing to look at further obscures pain and vulnerability through beautifying techniques.[45] And there may even be something bordering on the pornographic in looking when images appear to fulfill only the "prurient interests" of a viewer.[46]

6. Viewers

Clinical photographs look unmediated—as if they were made without human intervention. This quality makes them communicate as objective records of reality. But such representational rawness also gives the viewer the impression of having direct access to the subject on display. Material of this kind frequently showcases nudity, overt pathological symptoms, interoperative procedures, or other sensitive subjects. These are all "potentially visually upsetting" or "potentially disturbing" to viewers.[47] For the "realist" aesthetic of photography creates a sense of closeness that, in the case of medical images, makes both patient and disease appear as if physically present and temporally immediate. As Pichel outlines in her essay, the use of trigger warnings thus constitutes best practice in teaching with images of this kind. For it is not only those on display who need to be shielded: while the patient is beyond material harm, the viewer may be distressed by what they see.[48]

7. Indulgence

Photographs can make one feel deeply. But what is to be done with this response? In the words of Mieke Bal, the scattered range of emotions triggered by photographs of suffering often have "nowhere to go."[49] There is no way for the viewer to alleviate the distress on display. Such imagery "only shocks the senses; it does not teach meaning-making or historical truthfulness."[50] Even empathy and pity for the camera's victims can be unreflective if not self-indulgent, as these emotions allow little more than for those who look to revel in commiseration (to feel good about feeling bad for those on display).[51] There is nothing for those on display to gain from this encounter. Thus, to view these images is, at best, futile and, at worst, exploitative.

To Look or Not to Look?

With reasons both to look and not to look at clinical photographs, the historian is left with a dilemma.[52] Historical work has the responsibility of "rigorous use of evidence, invites counter-arguments, and seeks plausible, complex explanations about how and why past events, movements, concepts, and lives took place and shaped other past events, movements, concepts, and lives."[53] This desire to bear witness to the past is accompanied by the exposing, voyeuristic, and exploitative potentials of the photographic medium in medicine. Yet it is because of this tension that clinical photographs "raise important ethical questions of vulnerability, reproduction, and historians' complicity in the legacies of . . . unequal power relations."[54] As Pichel argues, it is thus not *whether* one looks but *how*. To echo Imada, what is needed is restraint—to query who benefits and who is harmed when one uses or does not use this material, and to ask *why* one is accessing, looking at, or reproducing these images.

NOTES

1. The broader field of "medical photography" includes asylum photographs (images of patients produced in psychiatric institutions), microphotography (made through the microscope), specimen photographs (showing disembodied human organs), surgical photographs (taken during surgical procedures), and radiographs (made with x-rays), as well as epidemic photographs (capturing the social experience of epidemics).

2. Mieneke te Hennepe, "Private Portraits or Suffering on Stage: Curating Clinical Photographic Collections in the Museum Context," *Science Museums and Research* 5 (2016), https://dx.doi.org/10.15180/160503; Mike Sappol, "Anatomy's Photography: Objectivity, Showmanship and the Reinvention of the Anatomical Image 1960–1950," *REMEDIA* (2017), http://www.diva-portal.org/smash/get/diva2:1188722/FULLTEXT01.pdf.

3. Rochelle Keene and Julie Parle, "Museums, Archives and Medical Material in South Africa: Some Ethical Considerations," *South African Museums Association Bulletin* 37 (2015): 12.

4. Jeffrey Mifflin, "Visual Archives in Perspective: Enlarging on Historical Medical Photographs," *American Archivist* 70 (2007): 32–69.

5. Te Hennepe, "Private Portraits."

6. Julie Parle, "The Voice of History? Archives, Ethics and Historians" (paper delivered at History and African Studies Seminar, University of Kwazulu-Natal Durban, 2005), 12.

7. See Katherine Rawling, "Visualising Mental Illness: Gender, Medicine and Visual Media c. 1850–1910" (PhD diss., Royal Holloway College, University of London, 2011); Susan Sidlauskas, "Inventing the Medical Portrait: Photography at the 'Benevolent Asylum' of Holloway, c. 1885–1889," *Medical Humanities* 39 (2013): 29–37; Rory du Plessis, "Photographs from the Grahamstown Lunatic Asylum, South Africa, 1890–1907," *Social Dynamics* 40, no. 1 (2014): 12–42; and Rory du Plessis,

"Beyond a Clinical Narrative: Casebook Photographs from the Grahamstown Lunatic Asylum, c. 1890s," *Critical Arts* 29, no. 1 (2015): 88–103.

8. Frank Möller, "Rwanda Revisualized: Genocide, Photography, and the Era of the Witness," *Alternatives* 35, no. 2 (2010): 130; Sappol, "Anatomy's Photography," 8.

9. Caroline Bressey, "The City of Others: Photographs from the City of London Asylum Archive," *Interdisciplinary Studies in the Long Nineteenth Century* 13 (2011), https://doi.org/10.16995/ntn.625.

10. Parle, "Voice of History?"; Suzannah Biernoff, "Medical Archives and Digital Culture," *Photographies* 5, no. 2 (2012): 179–202.

11. Te Hennepe, "Private Portraits."

12. Te Hennepe.

13. Ilze Sirmā and Leva Lībiete, "Blindspots: Unexpected Emotions Caused by Amputees in a Medical Museum," in *Beyond the Museum Walls: Medical Collections and Medical Museums in the 21st Century*, ed. Alfons Zarzoso (Barcelona: EAMHMS, SCHCT-IEC, and Museu d'Història de la Medicina, 2018), 140–141.

14. Erica Lehrer, Cynthia Milton, and Monica Patterson, *Curating Difficult Knowledge: Violent Pasts in Public Places* (New York: Palgrave Macmillan, 2011); Keene and Parle, "Museums, Archives"; Susan Lawrence, *Privacy and the Past: Research, Law, Archives, Ethics* (New Brunswick, NJ: Rutgers University Press, 2016).

15. For broader discussions on the use of photographs of various historical atrocities, see Frances Guerin and Roger Hallas, *The Image and the Witness: Trauma, Memory, and Visual Culture* (New York: Wallflower, 2007); Mark Reinhardt, "Picturing Violence: Aesthetics and the Anxiety of Critique," in *Beautiful Suffering: Photography and the Traffic in Pain*, ed. Mark Reinhardt, Holly Edwards, and Erina Duganne (Chicago: University of Chicago Press, 2006), 13–36; Ariella Azoulay, *The Civil Contract of Photography* (New York: Zone Books, 2008); Susie Linfield, *The Cruel Radiance: Photography and Political Violence* (Chicago: University of Chicago Press, 2010); and Mark Reinhardt, "Painful Photographs: From the Ethics of Spectatorship to Visual Politics," in *Ethics and Images of Pain*, ed. Asbjørn Grønstad and Henrik Gustafsson (New York: Routledge, 2012), 33–56.

16. Paul Weindling, *John W. Thompson: Psychiatrist in the Shadow of the Holocaust* (Rochester, NY: University of Rochester Press; Suffolk, UK: Boydell and Brewe, 2010); Paul Weindling, "'Cleansing' Anatomical Collections: The Politics of Removing Specimens from German Anatomical and Medical Collections 1988–92," *Annals of Anatomy* 194, no. 3 (2012): 237–242; Paul Weindling, "Introduction: A New Historiography of the Nazi Medical Experiments and Coerced Research," in *From Clinic to Concentration Camp: Reassessing Nazi Medical and Racial Research, 1933–1945*, ed. Paul Weindling (London: Routledge, 2017), 3–32.

17. Rebecca Onion, "History, or Just Horror? Should Archives Make Images of Eradicated Diseases and Antiquated Treatments Available for

the World to See?," *Slate*, November 6, 2014, https://www.slate.com
/articles/news_and_politics/history/2014/11/old_medical_photographs
_are_images_of_syphilis_and_tuberculosis_patients.html.

18. This is also the case for depictions of First Nations Australians.
Here, colonial photographs have made it possible to enact "native title"
claims for land historically stolen from Indigenous communities. See
Gaynor Macdonald, "Photos in Wiradjuri Biscuit Tins: Negotiating Relatedness and Validating Colonial Histories," *Oceania* 73, no. 4 (2003): 225–242;
and Michael Aird, "Developments in the Repatriation of Human Remains
and Other Cultural Items in Queensland, Australia," in *The Dead and Their
Possessions: Repatriation in Principle, Policy and Practice*, ed. Cressida Fforde,
Jane Hubert, and Paul Turnbull (New York: Routledge, 2002), 303–311.

19. Randall Jimerson, "Archives for All: Professional Responsibility
and Social Justice," *American Archivist* 70, no. 2 (2007): 253. See also
Patrick Ngulube, "Future of the Past: Access to Public Archives in
Southern Africa and Challenges to Historical Research," *Historia* 47,
no. 2 (2018): 562–582.

20. Martin Legassick and Ciraaj Rassool, *Skeletons in the Cupboard:
South African Museums and the Trade in Human Remains, 1907–1917* (Cape
Town: South African Museum, 2000), 1.

21. Susan Sontag, *Regarding the Pain of Others* (New York: Picador,
2003): 114.

22. Stephen Kenny, "Capturing Racial Pathology: American Medical
Photography in the Era of Jim Crow," *American Journal of Public Health*
110, no. 1 (2020): 76.

23. Judith Butler, *Frames of War: When Is Life Grievable?* (London:
Verso, 2009), 96.

24. Tina Campt, *Image Matters: Archive, Photography, and the African
Diaspora in Europe* (Durham, NC: Duke University Press, 2012), 16;
Linfield, *Cruel Radiance*, 29.

25. Biernoff, "Medical Archives," 196.

26. Glen Ncube, "Picturing Colonial Medicine: Perspectives on Patient
Photography" (paper delivered at Archives and Public Culture Research
Initiative seminar, University of Cape Town, October 18, 2012); Michaela
Clark, "Patients, Power and Representation: Clinical Photographs in
Focus," *De Arte* 54, no. 3 (2019): 59–76.

27. Luis Salazar, Amaya Maruri, and David Aranda, "Approaching
Emotion: Olavide Museum's Wax Moulages as an Element of Transformation," in Zarzoso, *Beyond the Museum Walls*, 135.

28. Similarly, photographs from apartheid-era passbooks used to
restrict the movements of people of color have been enlarged, colorfully
painted, and reworked as family portraits. See John Peffer, "Vernacular
Recollections and Popular Photography in South Africa," in *The African
Photographic Archive: Research and Curatorial Strategies*, ed. Christopher
Morton and Darren Newbury (London: Bloomsbury, 2015), 115–134.
Australian slave imagery is harnessed by Indigenous activists as a form

of public protest. See Jane Lydon, "'Behold the Tears': Photography as Colonial Witness," *History of Photography* 34, no. 3 (2010): 234–250.

29. Jason Bate, *Photography in the Great War: The Ethics of Emerging Medical Collections from the Great War* (London: Bloomsbury Academic, 2021).

30. Bate.

31. Jennifer Schuessler, "Your Ancestors Were Slaves. Who Owns the Photos of Them?," *New York Times*, March 22, 2019, https://www.nytimes .com/2019/03/22/arts/slave-photos-harvard-lawsuit.html.

32. Ian Berle, "Clinical Photography and Patient Rights: The Need for Orthopraxy," *Journal of Medical Ethics* 34, no. 2 (2008): 89–92; Te Hennepe, "Private Portraits."

33. Matthew Harting et al., "Medical Photography: Current Technology, Evolving Issues and Legal Perspectives," *International Journal of Clinical Practice* 69, no. 4 (2015): 401–409.

34. Lawrence, *Privacy and the Past.*

35. Robert Bogdan, "The Social Construction of Freaks," in *Freakery: Cultural Spectacles of the Extraordinary Body*, ed. Rosemarie Garland-Thomson (New York: New York University Press, 1996), 29.

36. Rosemarie Garland-Thomson, "Introduction: From Wonder to Error—a Genealogy of Freak Discourse in Modernity," in Garland-Thompson, *Freakery*, 4.

37. Examples include photographs produced in state facilities or during colonization, times of war, or economic depression. The existing literature is too great to cite, but for a vital overview, see John Tagg, *The Burden of Representation: Essays on Photographies and Histories* (Minneapolis: University of Minnesota Press, 1988).

38. Abigail Solomon-Godeau, *Photography at the Dock: Essays on Photographic History Institutions and Practices* (Minneapolis: University of Minnesota Press, 1991).

39. Jane Nicholas, "A Debt to the Dead? Ethics, Photography, History and the Study of Freakery," *Social History* 47 (2014): 139–155.

40. Biernoff, "Medical Archives," 189.

41. Berle, "Clinical Photography," 90.

42. Nicholas, "Debt to the Dead?"

43. Laura Mulvey, "Visual Pleasure and Narrative Cinema," in *Film Theory and Criticism: Introductory Readings*, 7th ed., ed. Leo Braudy and Marshall Cohen (1975; New York: Oxford University Press, 2009), 713.

44. Graham Clarke, *The Photograph* (Oxford: Oxford University Press, 1997), 133.

45. Susan Sontag, *On Photography* (New York: Farrar, Straus and Giroux, 1973).

46. Carolyn Dean, "Empathy, Pornography, and Suffering," *Differences* 14, no. 1 (2003): 88–124; Susan Crane, "Choosing Not to Look: Representation, Repatriation, and Holocaust Atrocity Photography," *History and Theory* 47, no. 3 (2008): 309–330.

47. Helen Wakely and Carly Dakin, "Assessing the Sensitivity of Images in Research Collections: A New Approach at the Wellcome Library," *Journal of Visual Communication in Medicine* 38, no. 1–2 (2015): 55, 57.

48. For more on this, see Marianne Hirsch, "Surviving Images: Holocaust Photographs and the Work of Postmemory," *Yale Journal of Criticism* 14, no. 1 (2001): 5–37; Guerin and Hallas, *Image and the Witness*; Linfield, *Cruel Radiance*; and Möller, "Rwanda Revisualized."

49. Mieke Bal, "The Pain of Images," in Reinhardt, Edwards, and Duganne, *Beautiful Suffering*, 93.

50. Crane, "Choosing Not to Look," 316.

51. Bal, "Pain of Images"; Crane, "Choosing Not to Look."

52. Frank Möller, "The Looking/Not Looking Dilemma," *Review of International Studies* 35, no. 4 (2009): 781–794.

53. Lawrence, *Privacy and the Past*, 110.

54. Nicholas, "Debt to the Dead?," 141.

CHAPTER TWENTY

Citation

COURTNEY E. THOMPSON

Citation is celebration
   ceremony
   community
   memory
   acknowledgment
   reproduction
   performance
   property
   erasure
   exclusion
   violence
   power[1]

I am a habitual overciter. This feature of my writing has frequently been pointed out and gently mocked in workshops and conversations. And yet I confess that my citation obsession has not always been informed by a critical lens as to how and why I (or others) cite. I have long treated it as a necessary—and anxiety-inducing—stage in the writing process, one that was mostly about due diligence, a way to overcome imposter syndrome, and a response to fears of accusations of plagiarism, as Susan Reverby suggests in the conclusion to this volume. Recent conversations on citational justice, however, have prompted me to rethink my approach to citation. What do citation practices *really* say about us as authors? What do they signal to our readers about us as scholars, about our interests, our theoretical and methodological approaches, our political orientations, our belief systems? And how can we best cite ethically?

For historians of science and medicine, there is a particular irony to uncritical citation practices. In the 1970s, the study of patterns and practices of citation became a central analytical tool in the sociology of science. Historians and sociologists of science studied citational patterns (and continue to do so) to evaluate the influence of particular scientists, laboratories, or schools of thought. While our allied fields have experienced various methodological turns since this point, it is suggestive that the analysis of the influence of citations of *scientists* was foundational to many approaches, but meta-analysis of the influence of *historians* of science has not captured the same interest. Historians of science, medicine, and health should turn the same critical lens on their own practices, considering how they signal their concerns, communities, and other influences on their work through citation.

In this essay, I explore some of the contours of this debate, inspired by the work of Black feminist and Indigenous scholars who are leading the discussion on citational ethics and justice. I want to prompt the reader to consider who and why we cite, as well as what our citations signal about our affiliations, our values, and our communities. I also want us to rethink the process of *not* citing, of refusing membership to those with whose work or personalities we do not wish to be in community. Citation is—and reveals and replicates—power, and we need to be more thoughtful in how we wield it.

Citational Justice

Not long ago, I was refereeing an article for a journal when I noticed that the author had only cited the work of white, male scholars, with few exceptions. The author had neglected the many women and people of color who had published on the topic over the past decade, citing only two women, one long dead (the other was me, which is probably why I was asked to review). I noted this in my review of the essay, stating that I could not deem such an article to be acceptable for publication unless the author cited and engaged with the work of scholars of color and more female scholars. (The article was eventually published by a different journal, with a few more female scholars cited.)

This erasure of the work of women, people of color, or people of other marginalized identities is a common enough occurrence that recent activism has addressed this phenomenon as one of the kinds of "silence" that Ayah Nuriddin explores in this volume. Most significantly,

#CiteBlackWomen emerged in 2021, in the wake of Black Lives Matter. The Cite Black Women Collective, an interdisciplinary group led by Christen A. Smith, drew inspiration from the Combahee River Collective, demanding both "reparations around citational practices" and "joyous moments" through this praxis.[2] "Black women have been systematically unnamed. Therefore, naming is an act of rebellion," the authors of the collective statement declared, making it clear that "Cite Black Women is more than just a catchphrase or a hashtag: it is an emphatic statement, a command, a rebuke, a call to action, a celebration, an act of rebellion, an ethos, and an act of love."[3]

#CiteBlackWomen inspired a wave of writing on citational politics and citational justice, especially around the citation of marginalized scholars. But in addition to rethinking *whom* we cite on the individual level, we can also consider citation as a community-building exercise, moving away from Western hegemonic norms of citation as property or performance. Māori scholars Hana Burgess, Donna Cormack, and Papaarangi Reid, following the work of scholars including Linda Tuhiwai Smith, acknowledge how citational practices reproduce the power dynamics at the heart of settler colonialism. They imagine a citation practice based on Māori principles instead, one that privileges collectivity and signals intergenerational relationships: "Through whakapapa [intergenerational networks of relationships], citation becomes a process of acknowledging the relationships through which knowledge emerges. It is acknowledging that our knowledge is not ours alone but is occurring as a part of a whakapapa. To 'Cite' is to call forth. In citing, we are calling forth past and future generations."[4] These authors call instead for balance, considering "accountabilities and consequences" that result from "calling forth" other authors and their ideas in our own work.

This is also a model of citation that centers notions of futurity— looking forward, to the worlds we create with our research, rather than simply backward, at that which has come before. As Burgess, Cormack, and Reid argue, "Our citation lists should read as the futures we are envisioning."[5] This principle fits well with the ethos of #CiteBlackWomen: the citation list should indicate the scholarly community one imagines, not just the so-called important works and scholars, a list that tends to prioritize white, male, and Eurocentric scholarship.

If one imagines a future field where the words and ideas of nonwhite and non–cisgender male scholars are read and valued, this should be reflected in our bibliographies. Moreover, this means not reflexively defaulting to the "important" thinkers of the past but considering the ideas we *intentionally* choose to shape scholarly (and other) futures. For Burgess, Cormack, and Reid, the ethics of a research program is a central question that must be interrogated, and that transformation (of the field, of the future, of the world) is a key component to ethical research. This is one of the ways to combat the influence of settler colonialism, they argue.

Two further points on inclusive citation require some thought. First, we must consider the limits of inclusive citation due to our own limitations as scholars. Our ability to access scholarship is often determined by our language competencies: we can only cite those books and articles that we can read. The seeming primacy of English for historical scholarship means that scholarship in other languages is often ignored or overlooked, or simply cannot be accessed by those with limited language skills. Moreover, the Western-centric nature of scholarship also gives priority to scholarship written by those in the Global North; scholars are expected to cite the "major works" written in English by European and American scholars, but they are not expected to be conversant with the scholarship written by scholars of the Global South or non-English-language scholarship. What this means is that a scholar could write a book on, for example, public health in India, without substantially reading or citing works written by Indian scholars or in languages other than English (or the few European languages deemed essential to scholarship, such as German and French), or without even setting foot in Indian archives. Scholars of the Global South are not accorded an inverse leeway in their uses of sources, whether archival or historiographic; they are expected to engage primarily with scholarship written by scholars of the Global North and are harshly criticized when they fail to do so. So-called global conferences, organizations, and journals also prioritize Western participants, another failure of accessibility, as gestured to by Nicole Lee Schroeder in this volume.

Second, we should consider what we can learn from those outside the scholarly community, and we should also cite the work of nonscholars, including descendant communities, activists, and patients, as suggested by Antoine Johnson in this volume with regard to his work with activist Gloria Lockett on the California Prostitutes Education Project. We should more seriously consider the non-peer-reviewed work of scholars, such as digital scholarship or even informal work in mediums like X (formerly Twitter). We must consider citation as praxis as we engage with our primary sources, especially those that capture the voices of marginalized peoples, including patients. While engagement with these sources is addressed by other authors in this volume, we should also consider how we cite our primary sources—how we include these voices and frame them as *actors,* not just subjects. Can citation also help us to think inclusively about the cocreation of knowledge *with* communities and patients of the past and present?

On *Not* Citing

A few years ago, I was chatting with a colleague who was an editor for a scholarly journal, and we argued about the politics of review (and citation) of scholars who have done wrong—not necessarily (just) as scholars but as people in the world. The short version is that his journal was publishing a review of a book written by a man who was then serving time for statutory rape. I argued that the review should not be published; the editor disagreed. He argued that if the work was good, it should stand on its own, and that if the work was valuable, it should be reviewed and disseminated—death of the author, in sum. I argued that reviewing the book at all without even adding a note about the context, amounted to tacit approval of the author; at the very least, it would mislead scholars who were not aware of the rape conviction, who might go on to read and cite the book and increase the status of the author. I argued—and still believe—that the lives of actual human beings, and the violences done to them, should matter more than "the work."

I have had this argument with other scholars since this conversation, but this issue is of course not so simple, in part because of the spectrum of actions that might cause some to want to remove a scholar from their citations, and the personal nature of this decision. Sara Ahmed writes of her desire to not cite men, period.[6] I explicitly wish to not cite known abusers or those who have otherwise acted in horrible ways (white supremacists, for example). I would not willingly associate with a known white supremacist or abuser, so why would I want to be in scholarly community with them? Citation is a space to declare one's community; I intend to remove known abusers from mine, as much as possible.

This form of silencing or "canceling," as it might be called in 2025, is controversial. I have found, informally, that there is a clear gender distinction in the attitudes toward erasing known bad actors from our citations. When discussing these principles with cisgender men, I have experienced a great deal of pushback, even from those who I believe otherwise to be engaged in feminist and antiracist work. They ask, Shouldn't "the work" matter? Doesn't it go *too* far to omit bad actors from our bibliographies? And this suggests how resistance might manifest in processes of refereeing and review. (I repeat: no "work" should outweigh human life and suffering.)

But what of those authors whose work cannot be erased so easily from our bibliographies, whose work was in fact so foundational to a particular subfield that it feels impossible to remove them from our academic genealogies and bibliographies? For example, Holt Parker, classics professor at the University of Cincinnati, was arrested in 2016 and later convicted for possession of over a thousand video files of child sexual abuse materials. This arrest sent a shockwave through the classics community, not least because Parker was known for his foundational work on Roman and ancient sexuality. The ethics of citing his work have been well explored by classicists, a community of which I am not a part, but I have been thinking about this case ever since. It has formed the backbone of my refusal to cite known abusers. Not all classicists would agree with my stance: a quick look at Google Scholar reveals that Parker's work has been cited in hundreds of books and articles since his 2017 conviction.

But that is classics; what about health history? We surely have our own monsters lurking among us. One figure we might reconsider is Michel Foucault. (Foucault is not a historian, you might protest. This is true. But in the theoretical underpinnings of our field, it is hard to go far without tripping over Foucauldian notions and terminology. This very volume includes several uses of the concept of the medical or clinical gaze.) I have found Foucault's theoretical works to be essential to my own thinking; my first book explicitly engages with Foucault's *Discipline and Punish.* I was therefore shaken to read that Foucault was accused by Guy Sorman of the sexual abuse of children, specifically children of color, in 2021. While these accusations are not precisely new, they were to me at the time. There is ongoing debate as to the veracity of these accusations, but if Foucault's actions cannot be confirmed, for some these uncertainties about his attitudes and actions around child sex abuse are sufficient cause for pause. The burden of proof should not be on the voiceless victim. If the victim cannot speak, our silencing of the (possible) abuser is one way to balance the scales.

More recently, Sheila Jasanoff has been accused by former postdoc Claudia Gertraud Schwarz-Plaschg of abuses of power within Harvard's Science, Technology and Society Program, including minimizing and denying the sexual harassment Schwarz-Plaschg experienced from male colleagues. Schwarz-Plaschg was supported by others with experience and knowledge of Harvard's program; it is enough for me, certainly, to put Jasanoff on my noncitation list. Other scholars, such as Joy Lisi Rankin, have spoken out about the abuses they have experienced within their departments and the field. Yet for every scholar who speaks out, so many others nod in recognition but stay silent—often quite understandable, as our careers, stability, and health care often rely on staying silent, keeping our heads down, and bearing the weight of abusive behavior for the sake of sheer survival. Not everyone has the privilege to speak out or to "fire" their university, as Rankin did. Contingency and the needs of survival often muzzle marginalized scholars who bear the brunt of abuse and bad behavior.

But what about the cases that are *not* a part of public record or otherwise publicly addressed? What about scholars whose violations have been of Title IX or of principles of the Equal Employment Opportunity Commission, or those who are known to be "bad actors"? Where do you draw the line, especially when the bad actor is not widely known as such? Whisper networks can be valuable for knowing whom to stay away from at a conference reception, but it is harder to justify leaving a scholar out of a citation without something specific (and public) to point to. I do not want to cite a person who abuses their partner, their colleagues, their staff, or their students. But if these stories are only whispered, it can be hard to justify the decision to exclude some scholars from our genealogies of knowledge production. Thus, refusing to cite is not only a complex decision for writers, but it is something to which editors and peer reviewers must also be more sensitive—and of which they also must be more accepting. Declining to cite certain scholars should be possible without pushback, when it is based on stories that are not ours to tell.

Canceling is not quite the right word for this kind of selective omission or conscientious objection. *Silencing* might be more accurate. Whatever you call it, it is an uncomfortable concept for many scholars. Certainly I have said and done things I regret or that I know now to be wrong; I have made choices in past publications that I would not make today. To say a scholar should be erased from the record for one error is to ignore the possibility for people to grow and change, becoming more ethical actors, as bell hooks has observed in the epigraph of the introduction to this volume. Yet for each of us there must be a line at which the actions of the individual author should give us pause and should force us to consider whom we want to include and acknowledge in our work, with whom we want to share our community, intellectual and otherwise.

***

Theory is not and never has been my forte, and I felt at sea in embarking on this essay, having to engage not only with the complexities of thinkers like Jacques Derrida but also the disorienting critiques of citation as property, as an instrument of colonization and of white supremacy, as conformity. I was exposed to many ways of defining, interpreting, and critiquing the practice of citation, a practice I had never truly thought much about before. Not all of these arguments or critiques swayed me, but I was forced to think about my own values of citation—what it meant to me, why and how I cite, and what I signal about myself, my scholarship, and my point of view in so doing.

This essay is only a taste of the reading and thinking I have done, which I discuss more in my first note on this essay, and it is also only a first step on this path. I have more thinking to do about citation as praxis—as something I do, something I teach, and something I use (implicitly or explicitly) to signal (and perform) my affiliations and values. This kind of thinking must always be part of our approach to our scholarship, writing, and teaching. I am persuaded that citation has deep meaning and significance, though scholars may disagree as to precisely the interpretation of these practices. Citation is not neutral. And the path to explore the meanings of citation, wherever it leads you, must be taken for the sake of producing ethical work.

NOTES

1. In writing this essay, I necessarily considered my own approach to citation. I have elected to attempt something experimental, influenced by the structure of Annemarie Mol's *The Body Multiple: Ontology in Medical Practice* (Durham, NC: Duke University Press, 2003). This model makes it more explicit as to which authors most influenced this essay, as a sort of merging of an acknowledgments section and a historiographic essay. The epigraph to this essay is inspired by the language used across the works I have read, speaking to the various complex and contradictory meanings that citation holds for different authors and communities.

My starting point for thinking about these issues was an article that I read with the Consortium for History of Science, Technology and Medicine working group, Carrie Mott and Daniel Cockayne's "Citation Matters: Mobilizing the Politics of Citation toward a Practice of 'Conscientious Engagement,'" published in *Gender, Place and Culture* 24, no. 7 (2017): 954–973. Our conversation about this essay, and the contents itself, inspired me to assign myself this chapter.

The most useful collection of essays I read for this chapter were those published as a special issue of *Diacritics* (48, no. 3) in 2020 by Beverly Acha, Jacob Edmond, Marisa J. Fuentes, Chase Gregory, Vivian L. Huang, Annabel L. Kim, Jennifer C. Nash, and Matt Tierney. These essays, particularly those of Fuentes and Nash, pushed me to reconsider my assumptions about how and why we cite, and also complicated my understanding of the stakes in #CiteBlackWomen. Tierney's essay, which locates the politics of citation within the context of the adjunctification and casualization of the academic labor market, added a further dimension to my analysis. Teaching Citational Practice: Critical Feminist Approaches (https://journals.library .columbia.edu/index.php/citationalpractice/about) has been an essential guide to feminist practices and stakes of citation. The *Diacritics* special issue and this website are good starting points for anyone interested in learning more about critical, ethical citation practices.

The most useful introduction to #CiteBlackWomen is Christen A. Smith, Erica L. Williams, Imani A. Wadud, Whitney N. L. Pirtle, & The Cite Black Women Collective, "Cite Black Women: A Critical Praxis (A Statement)," *Feminist Anthropology* 2, no. 1 (2021): DOI:10.1002/fea2.12040. I was also influenced by Savannah Shange's "Citation as Ceremony: #SayHerName, #CiteBlackWomen, and the Practice of Reparative Enunciation," published in *Cultural Anthropology* 37, no. 2 (2022): 191–198. Building on these ideas, other scholars turned to citational practices in their own fields. I am inspired in particular by recent work by Travis Chi Wing Lau, "On Citational Justice" (https://medicalhealthhumanities.com /2024/02/08/on-citational-justice/) and "Notes toward an Ethics of Editorial Care" (https://medicalhealthhumanities.com/2022/02/07/notes -toward-an-ethics-of-editorial-care/), in *Synapsis*.

With regard to Indigenous methodologies, I have been most influenced by the work of Linda Tuhiwai Smith, especially her book *Decolonizing Methodologies: Research and Indigenous Peoples* (London: Zed Books, 2021). On the topic of citation, I drew heavily on Hana Burgess, Donna Cormack, and Papaarangi Reid's "Calling Forth Our Pasts, Citing Our Futures: An Envisioning of a Kaupapa Māori Citational Practice," published in *MAI Journal* 10, no. 1 (2021): 57–67. I have also found valuable the resources provided by the Civic Laboratory for Environmental Action Research, which addresses principles of decolonization, including citational politics.

On the uses of citational analysis in science and technology studies and related fields, a meta-analysis of this practice was particularly useful: Alan L. Porter's "Citation Analysis: Queries and Caveats," published in *Social Studies of Science* 7, no. 2 (1977): 257–267. This essay provided an overview of the practice and its uses for science and technology studies scholars, as well as critiques as to the limitations of this methodology.

Not citing is a tricky and controversial topic. Here, it was nonacademic writing (which we should be using and citing more in our work) that was most influential. Nicole Seymour's "Citation in the #MeTooEra" (https://edgeeffects.net/metoo-era-citation/) was influential on my thinking, as was Sarah Scullin's "Making a Monster" (https://edgeeffects.net/metoo-era-citation/), which directly addressed the Holt Parker case. Claudia Gertraud Schwarz-Plaschg's Medium essay "On Its 20th Anniversary, My Testimonial on the Harvard STS Program" (https://medium.com/@claudia_gertraud/on-its-20th-anniversary-my-testimonial-on-the-harvard-sts-program-64100f6caac7) was published as I was working on this essay and added a further dimension to my consideration. Joy Lisi Rankin's Medium essay "Why I'm Firing Michigan State: Sexual Harassment, Online Harassment, and Utter Institutional Failure," published in 2018 (https://medium.com/@drjoy/why-im-firing-michigan-state-sexual-harassment-online-harassment-and-utter-institutional-6663a6bde68e), has long shaped my thoughts on these matters. In November 2022, Rankin presented at a roundtable on sexual harassment at the meeting of the History of Science Society, which also included Samantha Muka, Frazier Benya, Jenna Tonn, and Donald Opitz. At this event, a number of questions were asked about what to do with the citation of known harassers, which motivated me to finish this essay. I tried to channel Rankin's maxim at the meeting: "Burn it all down."

An aspect of citation to which we should give more space is our "acknowledgments." Often the people we "acknowledge" rather than cite are the ones who most clearly shaped the ideas and form of one's writing. In this case, my essay was profoundly shaped by Kylie Smith, Jai Virdi, and Janet Golden, who offered incisive commentary that pushed me to defend my points and shape my arguments in more provocative ways. My key scholarly community during this process included these women, as well as Aparna Nair (who pushed me to think more about scholars of the Global South), Cornelia Lambert, Sharrona Pearl, Amanda Mahoney, Kelly

O'Donnell, Lauren MacIvor Thompson, Elizabeth Neswald, and Sarah Swedberg. If citation is the way we signal our scholarly community, these brilliant scholars are mine.

2. Christen A. Smith, Erica L. Williams, Imani A. Wadud, Whitney N. L. Pirtle, & The Cite Black Women Collective, "Cite Black Women: A Critical Praxis (A Statement)," *Feminist Anthropology* 2, no. 1 (2021): DOI:10.1002/fea2.12040, quotations on 2.

3. Smith et al., 1–2.

4. Hana Burgess, Donna Cormack, Papaarangi Reid, "Calling Forth Our Pasts, Citing Our Futures: An Envisioning of a Kaupapa Maori Citational Practice," *MAI Journal* 10, no. 1 (2021): 57–67, citation on 61.

5. Burgess et al., 64.

6. Sara Ahmed, *Living A Feminist Life* (Duke University Press: 2017), 15.

CHAPTER TWENTY-ONE

Peer Review and Editing

AHMED RAGAB

In 1820, Charles Babbage wrote a searing indictment of his fellow scientists. He argued that British science had fallen, as demonstrated by the fact that the number of publications had declined significantly compared with the increasing number of fellows. He argued that a new system was needed whereby fellows could be differentiated by their productivity.[1] Babbage's criticism had a significant impact worldwide, including in British colonies. Here was a key scholar discussing what he saw as a decline of British knowledge after the end of the Napoleonic Wars, just as the British empire was steadily expanding. Alex Csiszar explains that Babbage drew on a long history of privileging the character of scientists who are men of letters. Babbage's criteria excluded artisans, collaborators, and teams of scholars.[2] Science was a collective endeavor, but the glory went to the single author.

The pushback against Babbage's critique emphasized the importance of quality over quantity. This discussion paved the way for William Whewell's 1831 proposal that is often seen as the traditional birth moment of the Western scholarly peer-review system. Whewell proposed that the Royal Society request a report by several distinguished readers for every submission it received. These reports were to inform the council's decision on whether to publish a manuscript. The reports were to demonstrate the value of the submitted piece and facilitate its dissemination. The French Royal Academy had been implementing a similar system for almost a century, after receiving a royal decree that allowed publications approved by the academy to bypass censorship.[3]

It took more than a century for the term *peer review* to appear and spread.[4] Historians of Cold War science explain how peer review in American and Western European academia strengthened scientists' claims to self-governance.[5] While reviewers often were not peers—in many cases, they were more senior than the authors—peerness as a category emphasized scholars as a community that was capable of self-governance. Peerness provided a significant allusion (or illusion) to democracy, lack of centralization, or hierarchy, further emphasizing the claims of modern Western science to objectivity and neutrality.[6]

The history of peer review offers a reminder of the historical contingency of the system, its connection to particular academic and scientific structures, and its relatively short life. In starting this piece on ethics with a historical analysis, I argue that editing and peer review are dependent on specific structures rooted in the colonial history of modern Western academia. These structures create specific ethical and moral imperatives that are contingent on the historical development of the process at hand. In this sense, this chapter argues for historically contingent ethics that can only be realized through careful analysis of history. I argue that the colonial construction of academic knowledge provided both the impetus and the guardrails for our peer-review and editing systems and the ethics regime that governs them. Therefore, looking into the future requires a similarly historically relevant commitment to different structural organizations of knowledge production.

I start with a discussion of the history of gatekeeping as a function of the professionalization of knowledge. I then discuss the ethical systems that govern our current peer-review practices and conclude by considering developmental review as a model for knowledge production. The irony in this piece is hard to miss. This piece is a peer-reviewed chapter in a peer-reviewed volume, and it is authored by the editor of the book series in which the volume appears. In offering an analysis of peer review to be peer-reviewed, I am effectively creating a conflict of interest. I will argue in this piece that such ethical conundrums result from the negative exuberance of our current ethical regimes and that similar conflicts of interest appear often and repeatedly, especially in marginalized disciplines, such as histories of medicine in the premodern world and in non-Western contexts, owing to the relatively small size of these subfields.[7]

I.

The history of peer review in Western academia lies at the intersection of editing, improvement, and quality control. The reports composed for the French Royal Academy responded to a particular royal privilege—the capacity to bypass censors. This privilege implied that a measure of control needed to be imposed.[8] Whewell and one of his students, John William Lubbock, composed the first readers' report commissioned by the British Royal Society. Yet the two soon disagreed on what they planned to do. Whewell thought the report offered a chance to publicize the submitted memoir and possibly offer suggestions. Lubbock wrote a rather scathing report arguing that readers cannot stand idly by when a paper exhibits errors.[9]

Historically, gatekeeping is familiar to academics. It is a common practice within various professional groups as they undertake their journey of professionalization. In this process of professionalization, institutional voices are accentuated, qualifications are set and agreed on, and rules or standards are set to create an in-group and an out-group.[10] Setting up rules, with the attached systems of inclusion and exclusion, was also a way for professional groups to demonstrate the ability to self-govern. Cooks and pottery makers in the bustling marketplaces of medieval Cairo or Timbuktu selected profession leaders and developed and endured self-imposed rules to prove to market regulators that they could be trusted to stop fraud in their midst.[11] In this context, the symbolic value of peer review extends beyond maintaining the quality of scholarship to sustaining the professional integrity of modern Western academia.

Therefore, it is unsurprising that peer-review errors are often the most difficult to navigate. An article in *Hypatia* on transracialism that failed to engage with any scholarship on critical race studies elicited a massive wave of resignations from the journal's board. The resignations were not simply an objection to the article itself; they were a sign of a lack of confidence in the editors who failed to apply the "proper" standards of peer review. The article was peer-reviewed. But it was not peer-reviewed by scholars in critical race studies.[12]

In the same way, a non-peer-reviewed article arguing that vaccines caused autism became a touchstone for anti-vaxxer movements.[13] In this case, vaccine skeptics used academia's own processes against it to argue that the scientific elites are in the pockets of Big Pharma. Conversely, hoaxes, such as Alan Sokal's attempt to discredit the humanities, used loopholes in the peer-review process to prove that the humanities failed to apply proper standards of academic conduct.[14] In other words, if humanities scholars cannot engage in sufficient gatekeeping, they should not be entrusted to self-regulate.

This history demonstrates the complexity of the peer-review problem. On one hand, plagiarism or the falsification of data, among other academic cardinal sins, cannot be reliably prevented by peer review. On the other, peer-review quality control is, by definition, conservative: peer reviewers uphold existing structures and norms. This is not necessarily because they believe in these existing structures but because this is the nature of the assignment. Biases based on race, gender, sexuality, and socioeconomic class are well documented. Anonymized reviews are but a feeble attempt to solve this problem.

The sustaining life force behind peer review is not epistemological but socio-epistemic. Peer review has been offered to the public as an epistemological endeavor—a method for producing reliable knowledge. For example, in the discipline of history, the importance of this tool has been magnified. It is supposed to sift through lies and allow the sanctioned scholarship to tell the "real" story. The presumed presence of an ontological reality independent of our epistemology has always been history's claim of kinship to natural sciences. Peer review is thus held up as the tool to determine what is real. Yet the survival of this epistemological tool seems to have little to do with its epistemological efficacy. Instead, I argue that peer review is sustainable because of its socio-epistemic role—its capacity to maintain the coherence of a professional community. Peer review is the outward process by which modern Western scholars demonstrate their capacity to self-regulate and ensure clear standards for knowledge production. In this context, whether peer review provides accountability or is the best way to prevent fraud is beside the point as long as it continues to communicate this capacity to the community of scholars and those beyond it.

Peer review's socio-epistemic role prevents it from being ethics-neutral. On the one hand, it is tasked with judging authors' ethical commitments. On the other hand, it requires (or claims) to be performed through strict ethical rules as a critical service to the profession. Some of these ethics include anonymity, recusal in cases of conflict of interest, and secrecy. Here again, the effectiveness of the peer-review process cannot be measured solely through its epistemic effectiveness but rather through the combined role of organizing knowledge and organizing the community of scholars.

To be sure, these are not minor roles. Maintaining ethical and well-researched scholarship is paramount if we take our role seriously as scholars in education and knowledge making. Similarly, self-regulation is necessary, especially amid current attacks on academic freedom and critical scholarship. The challenge, however, is to think about tools that allow for innovative and critical knowledge making and maintain ethical knowledge production.

II.

It is easy to find documents and guides outlining best practices and ethical considerations for peer reviewers. Most significant publications and publishers have guides or borrow and refer to famous ones. For example, the Committee on Publication Ethics has a document outlining basic principles for ethical peer review.[15] The PLOS (Public Library of Science) guide to peer-review ethics is widely used in the United States.[16] The National Library of Medicine updated PLOS's recommendations, emphasizing the need for positive feedback.[17] The ethical regime that governs peer review, similar to other academic ethical regimes, can point to *The Belmont Report*, which was issued following the Tuskegee syphilis trials, as a critical moment in the production of ethical knowledge and knowledge about ethics.[18] This *Belmont* imprint can be observed in how the previously mentioned ethics statements or guides focus on confidentiality, bias, and conflicts of interest as crucial issues that reviewers must pay attention to. If the author is the research subject in this case, then positive and constructive feedback is an act of beneficence. Similarly, confidentiality, anonymity, and lack of conflict of interest guard against abuse of power, protecting the subject from potential harm.

While these rules and guidelines represent necessary guardrails, they remain within the realm of "negative exuberance." Jasbir Puar places negative exuberance within "a risk economy that attempts to ensure against future catastrophe." Negative exuberance is an abundant, intense activity set and directed in the negative. This means that such activity is not intended to produce positive knowledge but rather to prevent, guard, and protect—or, in Puar's terms, "ensure against future catastrophe." Built in response to the ethical catastrophe, the *Belmont* ethical regime is one of negative exuberance—it is meant to protect against abuses and prevent disasters but not necessarily to build new knowledge positively. In the same way, the ethical guidelines of peer review instruct reviewers on what not to do and help them avoid pitfalls that can lead to disasters.[19]

Constructions of negative exuberance are inherently conservative. Here, it is helpful to remember that the "catastrophe" of the Tuskegee syphilis trials, which prompted *The Belmont Report* and the ensuing ethical regime, was not a crisis of unethical behavior. Indeed, many similar studies, in detail and spirit, continued to be performed after the new ethical regime took effect. The catastrophe was that the study became public outside scientific and medical research halls, proving the community's failure to regulate itself. In this sense, the *Belmont* regime's negative exuberance bred systems of box-checking, language games, and narrative tools to protect the researcher from the protective regime—another step of negative exuberance. Almost any graduate student whose work would put them in front of an institutional review board (IRB) has received advice about how to phrase an IRB application properly to pass review. There is hardly a historian who has not heard complaints about the imposition of IRB approval on oral history, and any anthropologist can discuss how IRB processes are a waste of time. These observations are not meant to indict the scholars or the IRB process but rather to demonstrate the inefficiency of the process and its inability to ensure ethical behavior.[20]

The same applies to the ethics of peer review. While these ethical guides continue to proliferate and warn against what scholars should not do, the character of the mean or unethical reviewer haunts academia. The fact that almost every academic has a story about the notorious Reviewer 2 raises an important question: Who is Reviewer 2? Are there enough academics who do not complain and who can be reliably selected as Reviewer 2? Or is almost everyone a potential Reviewer 2 who has been or will be perceived as such at some point? We need more empirical data to answer these questions. Yet these unanswered questions prove how an ethical system of negative exuberance fails to achieve its goals. Many resources instruct us not to be mean, but precious few tell us how to be good.

III.

If the ethics of negative exuberance aim to protect against future catastrophes, positive ethics should aim at building new, more ethical realities. As explained earlier, peer-review goals remain to produce reliable knowledge and to ensure scholarly independence. A positive ethical framework builds toward these goals instead of protecting against the disasters that may happen as these goals are being achieved. More specifically, a developmental peer review accomplishes these goals differently and offers an opportunity to avoid the previously mentioned problems. The fact that peer reviewers stand today as judges on whether a piece can be published, published after significant revisions, or not published at all assumes that papers need to reach peer reviewers at a very advanced, if not entirely completed, state. The unicorn of publications is the one that passes peer review with no revisions required.

A developmental approach to editing and peer-reviewing means that pieces arrive at peer review as in-progress. The reviewers are not asked to "pass it" but rather to improve it through feedback. Similarly, authors are not looking at the peer-review process as a threshold to cross but rather as an instance of feedback from someone with a fresh perspective. This process mirrors how scholars produce knowledge—through drafting, workshopping, and feedback. In this view, conflict of interest becomes less of a concern. Friends, colleagues, and mentors are likelier to give me honest feedback because they are invested in my work. In this view, one could imagine a space where anonymity is unnecessary and authors thank peer reviewers by name, recognizing their roles in supporting a publication and providing it with needed feedback. This approach questions the mythology of the atomistic author. Instead, it offers a space for collective work.

This approach, however, raises a series of essential questions. Can we entirely rely on self-regulation? Will authors be ethically asked to withdraw a publication that has received so much feedback but does not seem to "get there"? Will this affect scholarly independence if peer review is not tasked with rejection and gatekeeping? A peer-reviewed piece, with the understanding that it would have been rejected if it were not up to standards, seems to withstand more political and public scrutiny. Will a developmental process undermine this role of peer review? I would argue that developmental editing and peer-reviewing must perform this regulatory role. While the rule should be development, reviewers and editors will still need to decline works that are significantly out of touch with scholarship, are unethical, or violate practices of scientific inquiry. At the same time, this system emphasizes the importance of training, including ethical training and education, in making scholarly knowledge. The space for regulating knowledge should not be at the doorstep of publication but rather in the classroom. Finally, a structure of positive ethics invites accountability to involved communities and guarantees scholarly independence through solidarity rather than acquiescence to political pressure.

Developmental peer review is ultimately an editorial process. This means it will involve a more magnified role for editors and publishers, who would guide such a process. It is easy to imagine how this magnified role can cause problems. Guarding against such issues should not be a negative but rather a positive approach—through increasing diversity and ensuring openness and transparency in the publishing process.

Conclusion

In her book *Our Bodies Belong to God*, Sherine Hamdy proposes a system of biomedical ethics that departs from the four principles or similar transcendent values to be rooted instead in the experience of patients and doctors.[21] Ethics become therefore intentionally malleable and iterative—changing and developing based on historical context and positionality. Following on Hamdy's proposition, I argue here for a contextual and iterative structure that changes, challenges, and moves with the circumstances of knowledge production.

Many scholars have explored the colonial roots of the current systems of knowledge production in contemporary Western academia.[22] These colonial roots do not only extend to or affect investigations of colonial history or of areas claimed "relevant" to colonialism and its history. The colonial accumulation of knowledge and wealth was crucial in the making and maintenance of modern Western knowledge regimes, and the product of this knowledge has often and continues to support, implicitly or explicitly, the continuation of colonial and discriminatory structures. Most recently, calls for politically engaged scholarship in classics, medieval studies, or Renaissance studies were faced with claims of academic neutrality and criticism of the imposition of political agendas on scholarship. These claims demonstrated a blissful ignorance of how Greek and Roman histories, for instance, are and were mobilized by fascists and extreme right-wing groups.

At one level, this shirking of responsibility is a political act that entails choices and consequences. At another level, the blissful ignorance itself is a product of the colonial and racial luxury afforded by the institution of Western academia. As a scholar of Islam, regardless of what I study, I have come across and had to answer many questions about extremism, "backwardness," and "conflict of civilizations." As a scholar from and working on countries that suffer from autocratic regimes or struggle under occupation or civil war, my work is scrutinized by these regimes, and I am expected to comprehend the political valence of my work. The blessed isolation of the ivory tower, limited as it might be, is a product of the colonial accumulation of wealth and knowledge.

This consideration provides an important historical context for discussing the ethics of publishing, editing, and peer review. The ethical imperative to provide more diverse and more equitable access to knowledge production is rooted in the historical contingency of colonial accumulation. To be sure, this historical contingency is not universally accepted. But the discussion of the ethical imperative for equality and diverse knowledge production must be analyzed or critiqued at the level of its historical context and not based on an appeal to some form of imagined universal principles.

In this chapter, I have argued for a structure of positive exuberance—an attempt at shifting our frame of thinking from what should not be done to what could and should be done. This paradigmatic shift does not solve all problems related to the ethics or mechanics of publishing, nor does it intervene in the economies of publication, uncompensated labor, and precarity of academic workers. Yet I propose to reorient the gaze of editing and peer review from a stance of protectiveness to one of development. As an iterative ethic built on its historical contingency, this proposal does not offer answers to every ethical question, nor does it prescribe specific behavior. Instead, it provides an alternative metric for judging ethical practice insofar as it contributes to development and growth.

As opposed to "kindness," a useful but amorphic descriptor, contribution to development is an intentional constructive measure. Here, the reviewers' reports are not only asked to be kind or nice but are rather tasked with being developmental and productive. The question, therefore, is whether a report or a review can contribute to making a project better and not whether a project is publishable. The resulting demystification and the declining importance of anonymity will also help provide clear professional rewards for review work in a manner like teaching and mentoring—a difficult prospect with the necessity of anonymity and the lack of clear metrics to judge good and bad reviews. To be sure, reviewers will reject projects or decline requests to mentor or develop them. Yet a developmental model emphasizes the capacity to help, support, and mentor—processes that are clear-eyed and committed to various ethical and structural regimes.

NOTES

1. Alex Csiszar, *The Scientific Journal: Authorship and the Politics of Knowledge in the Nineteenth Century* (Chicago: University of Chicago Press, 2018).

2. Alex Csiszar, "Peer Review: Troubled from the Start," *Nature* 532, no. 7599 (2016): 306–308.

3. Alex Csiszar, "How Lives Became Lists and Scientific Papers Became Data: Cataloguing Authorship during the Nineteenth Century," *British Journal for the History of Science* 50, no. 1 (2017): 23–60.

4. Adrian Johns, *The Nature of the Book: Print and Knowledge in the Making* (Chicago: University of Chicago Press, 2000).

5. See Andrew Jewett, *Science, Democracy, and the American University: From the Civil War to the Cold War* (Cambridge: Cambridge University Press, 2012); and Audra J. Wolfe, *Competing with the Soviets: Science, Technology, and the State in Cold War America* (Baltimore: Johns Hopkins University Press, 2013).

6. Paul Erickson et al., *How Reason Almost Lost Its Mind: The Strange Career of Cold War Rationality* (Chicago: University of Chicago Press, 2013).

7. In order to limit the multiple layers of potential conflict, this chapter was the only one anonymized during the review process. In a sense, it was still single anonymization but in the opposite direction—I, the author, knew who the reviewers were, but the reviewers did not know me. As shown later, I argue here that these forms of anonymization are a product of the current system and its negative exuberance. Here, it is important to acknowledge the professionalism and care with which the reviewers took on this challenge—reviewing a piece about reviewing. I am very grateful for their work and deep insights.

8. Csiszar, *Scientific Journal.*

9. Csiszar.

10. Michael D. Gordin, *The Pseudoscience Wars: Immanuel Velikovsky and the Birth of the Modern Fringe* (Chicago: University of Chicago Press, 2012).

11. See Peter E. Pormann, "The Physician and the Other: Images of the Charlatan in Medieval Islam," *Bulletin of the History of Medicine* 79, no. 2 (2005): 189–227; and Ann M. Blair, *Too Much to Know: Managing Scholarly Information before the Modern Age* (New Haven, CT: Yale University Press, 2010).

12. Rebecca Tuvel, "In Defense of Transracialism," *Hypatia* 32, no. 2 (2017): 263–278. For an overview of the controversy, see Jimmie Manning and Jennifer C. Dunn, "Rachel Dolezal, Transracialism, and the Hypatia Controversy: Difficult Conversations and the Need for Transgressing Feminist Discourses," in *Transgressing Feminist Theory and Discourse*, ed. Jennifer C. Dunn and Jimmie Manning (New York: Routledge, 2018), 3–15

13. Michael J. Smith et al., "Media Coverage of the Measles-Mumps-Rubella Vaccine and Autism Controversy and Its Relationship to MMR Immunization Rates in the United States," *Pediatrics* 121, no. 4 (2008): e836–e843.

14. Bob Hodge, "The Sokal 'Hoax': Some Implications for Science and Postmodernism," *Continuum* 13, no. 2 (1999): 255–269.

15. "Ethical Guidelines for Peer Reviewers," Committee on Publication Ethics, accessed November 21, 2024, https://publicationethics.org/resources /guidelines/cope-ethical-guidelines-peer-reviewers.

16. "Guidelines for Reviewers," PLOS ONE, accessed November 21, 2024, https://journals.plos.org/plosone/s/reviewer-guidelines.

17. *PLoS Medicine* Editors, "Peer Review in *PLoS Medicine*," *PLoS Medicine* 4, no. 1 (2007): e58.

18. Hiroyuki Nagai, Eisuke Nakazawa, and Akira Akabayashi, "The Creation of the Belmont Report and Its Effect on Ethical Principles: A Historical Study," *Monash Bioethics Review* 40, no. 2 (2022): 157–170.

19. Jasbir K. Puar, *Terrorist Assemblages: Homonationalism in Queer Times* (Durham, NC: Duke University Press, 2018), 8.

20. Rose McDermott and Peter K. Hatemi, "Ethics in Field Experimentation: A Call to Establish New Standards to Protect the Public from Unwanted Manipulation and Real Harms," *Proceedings of the National Academy of Sciences* 117, no. 48 (2020): 30014–30021.

21. Sherine Hamdy, *Our Bodies Belong to God: Organ Transplants, Islam, and the Struggle for Human Dignity in Egypt* (Berkeley: University of California Press, 2012).

22. Kwame Anthony Appiah, "Is the Post- in Postmodernism the Post- in Postcolonial?," *Critical Inquiry* 17, no. 2 (1991): 336–357; Saliha Belmessous, *Assimilation and Empire: Uniformity in French and British Colonies, 1541–1954* (Oxford: Oxford University Press, 2013).

CHAPTER TWENTY-TWO

Cultivating Ethical Strategies in Digital Health Humanities

BRITT DAHLBERG AND JESSICA MARTUCCI

Historians of medicine have been utilizing digital humanities (DH) tools to engage our imagined audiences for decades. An early example is Laurel Thatcher Ulrich's interactive website featuring the diary of the eighteenth-century midwife Martha Ballard.[1] Today, social media has expanded the possibilities for digital engagement, offering a crucial platform for public discourse where information is exchanged and spread rapidly to larger and larger audiences.[2] As health humanities scholars, what are our ethical obligations when writing in these hyperpublic spaces?[3]

We suggest there are specific ethical issues that arise when doing DH work. This work is not siloed within a single academic discourse; it inherently seeks to connect with broader discourses. Therefore, a key part of doing DH work is thinking critically about your intention: With whom do you want to be in conversation, and how do you want to relate to them? DH work raises questions about assumed divisions between researcher, subject, and audience.[4]

In the pages that follow, we do not prescribe one set of ethical rules for all future DH projects. In fact, we do not believe this is even ethically desirable. Instead, we argue that ethical DH work requires asking key questions and developing critical forms of attention throughout both the planning and doing of projects. Our goal is to help readers consider how they might build their own reflective ethical tools specific to their projects. To do this, we examine a project we developed in the midst of the COVID-19 pandemic to illustrate these processes of asking and acting on ethical questions.

The Beyond Better Project

Question 1: What are your ethical intentions, and how will these shape your project? We launched the Beyond Better Project in June 2020 at the beginning of the COVID-19 pandemic.[5] The dominant media and public health messaging in the United States at the time focused on narratives of "survival" or "death" while reassuring "healthy" Americans at the expense of the disabled. Our project challenged this ableist perspective. We aimed to explore the vast set of experiences that lay beyond the dichotomy of either dying or "getting better," by curating oral history interviews and artwork depicting firsthand accounts of epidemic experience, as well as by intervening in one realm of public discourse, Instagram. Our intentions required an embedded and relational ethics deeply informed by disability justice perspectives and the historical moment in which we and our participants lived.

Intentions drive research design. We interviewed narrators, curated their narratives, and exhibited edited text paired with illustrations from disabled artist-collaborators.[6] We selected oral history as our starting method because we aimed to collect and contextualize two epidemic diseases, COVID-19 and polio, within peoples' overall life stories in a way that centered their individual humanity and foregrounded the ways epidemics and disability are always intertwined. We chose Instagram as our exhibit space so we could connect our work with that of others in the disability justice/DH community in response to a historical moment of extreme isolation: writing to and for others living through the pandemic.

Ethical DH is not devoid of academic grounding; rather, our fields (history and anthropology) informed our design. Historical work on polio suggested that disability would be a key lens for examining the COVID-19 pandemic.[7] As we witnessed the public omission and silencing of disability experiences, the famous call from the disability rights movement, "Nothing about us without us," was ever present in our minds. In creating a DH project that directly engaged and intervened in public discourse, we drew from work in the disability studies and disability justice literatures because these were deeply needed perspectives informed by people with the most timely, relevant, embodied firsthand experience.

Imagined Audience

Question 2: When writing health care histories for "the public" online, whom are you writing to, for, *and* with? The history of medicine has been accused of uncritically embracing the perspectives of physicians and other scientific and medical experts.[8] The underlying assumption of this work was that it was written to practitioners, not people who were sick or disabled, and that reliable knowers were practitioners, not patients.[9] Although the turn toward patient perspectives and history "from below" has provided a partial remedy, the field still has much to do in order to embrace a more accurate and complete picture of health care and experiences of health, disability, and illness.[10]

Planning our *digital* health humanities project around the intersection of disability and health care brought these ethical concerns about subject and audience into sharp relief. Until recently, disabled people, when written about at all in the history of medicine, were too often treated and written *of* as objects of study and treatment, rather than *with* (recognized as creators of knowledge themselves) and *to* (as key audiences).[11] Doing *ethical* public DH work about peoples' experiences with illness and as patients required us to develop a collaborative methodology in which we understood that our participants would be our audience as well as our coanalysts. This prompted an ethical strategy that utilized a dialogic approach to writing histories of medicine in which we and our collaborators moved between roles of analyst, subject, and audience. By *dialogic,* we mean an approach in which knowledge, theory, writing, and artwork are created through exchange. Through engaging with us in oral history, art making, and the review of exhibits, our narrators and artists participated in cocreating knowledge, theory, and community with us. In the process, each of our collaborators, and the prior writers and activists we learned from, was helping us make sense of the moment we all were living through.[12]

Being conscious about audience and your own ethical commitments will ripple through each step of your work: research design, process, writing, and selection of a publication venue. The embedded, embodied, and relational ethics we developed for this project informed each stage of our "writing" via digital health humanities: selecting and interviewing narrator-collaborators, transcribing and curating words from that oral history exchange, discussing draft images with artist-collaborators, and sharing drafts with narrators as our audience and coanalysts.

Selecting Narrator-Audience-Collaborators

Question 3: Whom are you paying attention to, and how are you relating? Treating narrators as cocreators and audience members, not just subjects, meant we had to be especially selective in choosing whose stories we would tell. Conscious of the fact that we would be publishing intimate health information on a public platform, we sought narrators who had already chosen to share some of their health experiences online. This method led us to our first two narrators—female PhD holders with backgrounds in the history of medicine. These individuals brought a shared language and familiar ways of thinking about the past, gender, and disability, while also bringing their own angles and experiences. By interviewing our peers, we also reduced the power imbalances that typically loom over researcher-participant interactions, thus allowing us to pilot our experimental ethical strategy of collaborative cocreation that we hoped we could then carry forward into future projects with others.

Question 4: Whom are you doing right by? Or for whom are you seeking justice? We also chose to include transcripts from interviews done in the 1990s with people with post-polio syndrome by Britt's colleague, anthropologist-gerontologist Mark Luborsky. These sources provided historical perspective on the experience of epidemic diseases and their relationship to disability. One narrator's story especially captured our attention: the narrative of a woman we refer to as Anne. As an older adult, Anne became an advocate for those with post-polio syndrome, and she chose to participate in the interview for this reason. We thought critically about how or whether we ought to include this source from someone we would never meet. We spoke at length with Mark about his interviews and decided that including Anne's story felt in line with the intentions each of us had articulated. For our project, we sought to address silences by engaging with traditionally marginalized people on their experiences with health, illness, and disability and to situate these within a full life story. In the interview, Anne herself expressed her intention to share her experiences with others and participate in advocacy for polio survivors. Mark's intentions were to conduct life history interviews to explore firsthand polio experiences. All three intentions fit our ethical framework rooted in perspectives of disability justice and the centering of human experiences through personal historical narratives.

Implementing this ethical strategy meant that not all of our interviews made it through our process to the final exhibit. One interview, in particular, raised ethical flags. This narrator struggled to be open or reflective about their own embodied experiences and relationship to health, illness, and disability. Instead, they seemed to carry a predetermined political objective that did not align with the spirit of our project. We knew we could not share authorial power with this narrator, whose own ethics mismatched with ours and whose agenda felt at odds with the collaborative and disability-focused framework of our project. Staying true to your ethical strategy throughout your process may raise uncomfortable moments. Rather than derailing your work, these provide important opportunities for revisiting and honoring your ethical commitments.

Curating Text

Question 5: Whom are you creating for or with, and why? Selecting and editing from transcripts involved a continuous process of ethical deliberation. For us it was an opportunity to revisit *whom* we were creating this for—a question that kept us focused on the intention that this would be a positive and empowering experience for our participants and their communities. As Britt observed in her research notes while curating Deanna Day's interview, "In her oral history, [Deanna] weaves different threads. She raises something early on, heads to something else, and as interviewers we circle back later. But when editing, which storylines do I select? What expresses *her* as well as themes of the project?" The process of "editing down" is an active one, providing many opportunities for ethical reflection.

Guided by the desire to embrace our narrators as collaborators, as well as our future audience, we sought to create exhibits that conveyed the spirit of our narrators, the events and themes from their life stories that mattered to them in that moment, and the intended themes of our project. We used our memories of the interviews (what stuck with us over time), as well as transcripts, to help identify sections that expressed their goals and ours. We continued to move in a dialogic fashion: between our intentions and those of our narrators, and between recordings, transcripts, and curations, while seeking feedback from our narrators along the way.

To create artwork to accompany the texts, we brought in two artists who self-identified as disabled, Savannah Hall and Jade Ramos. We talked with them at length about the spirit and intention of the project, provided them with background readings, and introduced them to stories of our narrator-collaborators before asking them to create visual artworks inspired by sections of the interviews.

Creating visual art involved collaborative ethics and choices as well. We spent time discussing draft illustrations. For example, in our exhibit on the polio survivor Anne, she talks about having a new baby with her husband and figuring out how to navigate the stairs in their home by using one crutch with one arm "and holding the baby in the other like a football." In a first draft of the corresponding illustration, the artist sketched this scene from the vantage point of the top of the

Block letters, stenciled, and colored black, without spaces in between. The words wrap from one line to the next, filling up a square: "DEC WHAT'S HAPPENING IN WUHAN, CHINA?" Following the words are icons of TikTok, hand sanitizer, and a mask.
By artist Jade Ramos, 2021.

stairs, looking down on Anne. We realized, however, that this angle portrayed Anne's disability as a disadvantage in her motherhood, rather than conveying the meaning she expressed in her interview when she discussed the ways her life as a disabled woman had allowed her to develop and rely on her own ingenuity and creativity, abilities that she found helpful as a mother. After further discussion, the artist created what became the final piece for that collection. The illustration shows the same scene but from the perspective of looking up the stairs, with Anne in front and above the viewer—a perspective that felt far more reflective of Anne's experience as she had shared it, and one we would have felt comfortable sharing with Anne herself.

Thinking of narrators, sources, and artists as collaborators, not just content, meant that this *process* of exchange between each of us was a fundamental goal of our project. Later, the artist Savannah shared the ways that visiting with Jaipreet Virdi's story prompted insights and reflections on her own life:

> I felt her story deeply, feeling [in parts] how could this happen?! . . . I felt really connected to her . . . even though we don't share any of the same conditions, or in no point in our lives would I be in the same vicinity of her, but certain phrases and things she said, there is something I just connect with on a base level.[13]

Savannah's process was deeply relational, continuing the spirit of the project's methods as a space for asking, listening, sharing, and extending that process into her own network:

> I don't think I understood how emotionally draining it would be until I started watching the interview and reading the transcript. I'd pause and go tell my roommates, "I need to tell you what happens!"[14]

> Creating artwork like this allows you to generate a deeper kind of empathy because you're spending so much time and thinking deeply about the person behind the story. You're constantly thinking: okay, I need to create some way to portray this deeper meaning to other people at a glance, at one simple glance, I need them to absorb this [experience].[15]

Her considerations turned into choices about styles, effects, perspectives, framing, and color as she illustrated pieces of Jaipreet's and Anne's interviews.

An illustration of a woman on a staircase in a midcentury American home. She wears a blue skirt and blouse, her back toward us as she climbs the brown staircase, with a crutch under one arm and a baby tucked in the other. Family photos line the walls of the staircase.
By artist Savannah Hall, 2021. Original in color, reproduced in black and white.

Narrators as Audience: "Seeing My Story Reflected Back to Me"

Question 6: Is your work creating the desired effect? One part of narrator Deanna Day's story that caught our attention was a long series of difficult medical experiences in which her pain was repeatedly ignored or downplayed by medical professionals. She unfolded this story across various parts of our interview—after all, the interactions took place over many years, interspersed with all sorts of other things going on in her life. But as we worked on curating her exhibit, we chose to lay these out in one long narrative arc.

We sent the draft text to Deanna and discussed it by phone, open to hearing her feedback. She shared, "It was good to see that experience rendered plain. Because it is 15 years of soggy experiences, of pain, conversations, confusion. To see it all—not a timeline per se, but just laid out, as 'this happened.' And yeah! [Seeing it laid out, you see] that was really fucked up. But it felt illuminating to see it all together like that."[16] Hearing the ways this editing, curation, and sharing back affected Deanna heartens us about the capacity for DH projects like this to serve a role in offering spaces and methods for listening and being with the realities of a full human experience, particularly around illness and health. While we do not claim that this particular work is reparative at the community level, we do think that by providing a public platform for collaborative storytelling, DH projects can be.[17] In our case we found that this process helped our narrators and artists identify, make visible, and name a set of shared experiences with medical gaslighting, ableism, and trauma. Although we went into this undertaking concerned about the ethics of sharing intimate heath information online, we found instead a powerful imperative to lay out clearly embodied truths that had been repeatedly ignored in the medical system. In the end, this allowed us to not only *do less harm* but work toward a kind of individual repair that could then resonate and support others who might recognize these stories in their own experiences.

Building a Responsive Ethics

We have presented this project as a model for how to think through ethical decision-making in health-related DH projects and have argued that ethics in these projects need to be, and in fact are, embedded in every aspect of the process, from design to reception. Rather than being prescriptive, ethics in DH projects must be situated in relation to time, place, purpose, and the people involved—particularly those for whom and with whom you are writing. In working with individuals on their epidemic experiences in the context of their much longer lives, we made choices that aligned with our goals: to reject traditional hierarchies of knowledge making and disciplinary boundaries in favor of equalizing ourselves, our narrators, our artists, and our audience to create a space online where the challenges, trauma, and even beauty of the moment we were all living through could be on full display. By doing so, we hoped to offer a space for human connection, empowerment, and community to flourish at a time in which so many experienced immense disconnection, isolation, and powerlessness.

Digital history projects are wide and varied, so no one set of ethical considerations will be applicable across every project. What we have suggested here, then, are a series of questions that we have come back to again and again throughout the development, implementation, and afterlife of the Beyond Better Project:

1. What are your goals and intentions, and how should these shape your project design?
2. Whom are you seeking to be in dialogue with, why, and how?
3. Whom are you paying attention to, and how?
4. Whom are you trying to do right by?
5. Whom are you creating for or with, and why?
6. Is your work landing as you intend?

How you think through and act on these questions in your own work will guide you in forming an ethical framework and process that help you *do less harm.*

NOTES

1. Laurel Thatcher Ulrich, *A Midwife's Tale: The Life of Martha Ballard, Based on Her Diary, 1785–1812* (New York: Vintage, 1991). DoHistory, the website developed to accompany Ulrich's book, is still online at https://dohistory.org/ (accessed July 25, 2023).

2. Some of the social media movements we are most familiar with include #twitterstorians, #histmed, and #disabledinSTEM.

3. We use the term *health* humanities as opposed to *medical* humanities because we are interested in the full spectrum of human health knowledge and experiences, including those of patients, nurses, and other health care workers; *health humanities* centers intersectionality and interdisciplinarity; and it emphasizes social and cultural explanations for and understandings of disease and disability. For more on this comparison, see Therese Jones et al., "The Almost Right Word: The Move from *Medical* to *Health* Humanities," *Academic Medicine* 92, no. 7 (July 2017): 932–935.

4. Though we focus on DH, we invite readers to explore ways in which these points extend into their academic work as well.

5. We thank the National Council on Public History for a programming grant that helped us launch this project. We also thank the Albert

Lepage Center for History in the Public Interest at Villanova University for its COVID-19 Grant for History in the Public Interest that supported our artist-collaborators, as well as a further internship grant that paid student interns and trainees who worked with us.

6. The exhibit can be found, as of July 2023, on Instagram (@thebeyond betterproject) and at https://www.beyondbetter.org.

7. See, for example, Daniel J. Wilson, *Living with Polio: The Epidemic and Its Survivors* (Chicago: University of Chicago Press, 2005).

8. For a critique of this past from a critical race theory perspective, see the work of Antoine S. Johnson, Elise A. Mitchell, and Ayah Nuriddin, "Syllabus: A History of Anti-Black Racism in Medicine," *Black Perspectives* (blog), August 12, 2020, https://www.aaihs.org/syllabus-a-history-of-anti -black-racism-in-medicine/.

9. See Nicole Schroeder's essay in this volume.

10. Roy Porter, "The Patient's View: Doing Medical History from Below," *Theory and Society* 14, no. 2 (1985): 175–198.

11. Beth Linker, "On the Borderland of Medical and Disability History: A Survey of the Fields," *Bulletin of the History of Medicine* 87, no. 4 (Winter 2013): 499–535.

12. We are particularly indebted to the work of disability justice leaders Alice Wong and Mia Mingus, as well as the academic work of Daniel J. Wilson.

13. Savannah Hall, debrief with Jessica Martucci and Britt Dahlberg, May 28, 2021.

14. Hall.

15. Hall.

16. Deanna Day, debrief with Britt Dahlberg, November 5, 2021.

17. See Kylie M. Smith's essay in this volume.

PART V

Teaching

CHAPTER TWENTY-THREE

Centering the Margins

ANTOINE S. JOHNSON

My first time teaching the Introduction to the History of Medicine course resulted in fireworks. The first sentence in the course description noted that students would examine ways in which racism is foundational to American medicine. In doing so, students read material that centered the experiences of minoritized ethnicities—primarily African Americans—to understand ways in which medical racism affected and continues to affect these groups. Many students enrolled in the course were mid- to late-career physicians, eager to earn a master's degree in the history of medicine. As such, they did not anticipate ongoing conversations about ways in which racism pervades American institutions, specifically American medicine. This resulted in several class discussions that turned into shouting matches, as some physicians assumed I had called them racist as opposed to highlighting its influence in the field of medicine. This took away some rare opportunities to have critical dialogue about problems with, and inside, American institutions. Some even thought it was inappropriate for a non-MD historian of medicine to teach the history of medicine, comparing it to a man instructing a woman in what she can and cannot do with her body.

The pushback I received in class convinced me to discuss ethics openly in the context of my positionality, as well as how I decided what to include in the syllabus. This led to historical and contemporary examples of why centering minoritized groups can be an effective approach in the history of medicine and health care. That included the ongoing coronavirus (COVID-19) pandemic, which has only illuminated the extent to which Black people look askance at the medical community. For example, in December 2020, while speaking to a group of Black health professionals, scholars, and religious leaders, Anthony Fauci of the National Institute of Health urged African Americans to "put skepticism aside" and get the coronavirus vaccine. Fauci acknowledged African Americans' hesitance, citing the US Public Health Service's syphilis experiment at Tuskegee, where Black men were monitored—not treated—for syphilis from 1932 to 1972.[1] Despite that history, Fauci argued it "would be doubly tragic that the lingering effects of that prevent you from doing something so important."[2]

Providing the vaccine alone, without addressing the social and environmental conditions that left—and continue to leave—African Americans susceptible to infection, is like administering a bandage for a bullet wound. Less than a year into the pandemic, African Americans were three times more likely to die from coronavirus infection than white Americans.[3] Few people expressed this more clearly than retired football player Marshawn Lynch. In a conversation with Lynch, Fauci mentioned that upon infection, African Americans have "a much greater chance at getting seriously ill and dying than whites do." Lynch immediately responded by asking what it is that causes diseases to take a more "strenuous toll" on Black people than on their white counterparts. In response, Fauci highlighted decades, if not centuries, of structural violence leading to illnesses—diabetes, hypertension, kidney disease, chronic lung disease, and obesity—that will produce worse outcomes for African Americans. Fauci's response left much to be desired, giving the impression that African Americans were biologically different and thus more vulnerable to infection. As such, the conditions he cited were larger issues that the coronavirus vaccine alone could not fix. As Lynch responded, "We still haven't tackled the problems [that] leave us more susceptible."[4]

Lynch's sentiments demonstrate the importance of not only decentering whiteness but also focusing less on biomedical developments and more on culturally specific outreach to address health inequities. The early AIDS epidemic illustrated the importance of such an approach, as AIDS was first portrayed as a "gay white disease." This did irreparable harm for African American AIDS activists who struggled to convince poor and working-class Black people to take the epidemic seriously. In 1987, Wesley Anderson—cofounder of Philadelphia's Blacks Educating Blacks about Sexual Health Issues—lamented that "prominent articles about AIDS in *Time, Newsweek,* and on network [television] never showed any black AIDS victims or health professionals."[5] These actions reinforced misconceptions of AIDS as a "white disease," exacerbated by inequitable resource distribution and awareness campaigns catering largely to the needs of white men.

Even before encountering students in my Introduction to the History of Medicine class, I felt a calling to amplify the experiences of minoritized groups, using antiracist pedagogy as an ethical stance in the field. Some of my students wanted to learn the field from what I call the three-Ds perspective—doctors, drugs, diseases—that privileges and perpetuates "great white men," with little effort made to understand the field from a critical medical humanities perspective.[6] Through primary and secondary sources, and aided by an explicit engagement with racism and race in medicine, students enrolled in history of medicine and health care courses gain, from pedagogies similar to mine, perspectives on ways in which racism is foundational to American medicine. By extension, this approach challenges learners to interpret the documents from alternative viewpoints. Using personal examples and the archives of the experiences of patients, activists, and everyday folks, courses and research projects are designed to complicate preconceived notions of the impact racism has on one's health.

Many scholars put on our syllabi what we have read independently, as opposed to things assigned in courses we took as students. But some of us run into the problem of archival silences, creating questions about what happens when marginalized groups are not accounted for. How, then, do their stories get told ethically? And what does that say about the archives, public opinion, and public policy affecting said communities? These are some questions that resonated while conducting research on the California Prostitutes Education Project (CAL-PEP). Founded in 1984 by Gloria Lockett in San Francisco, CAL-PEP fought rigorously to reduce new AIDS cases among sex workers and eradicate stigmatization and scapegoating of women in the industry, and it partnered with politically aligned health professionals to provide resources to poor and working-class Black and Latinx folks throughout the Bay Area. In the early 1980s, after a heterosexual man tested positive for AIDS, "public-health officials assumed these results were due to their interactions with prostitutes, representing the historically familiar pattern of scapegoating prostitutes for the spread of disease."[7]

It was important for me to share my research on CAL-PEP with my students, many of whom were physicians who saw patients daily. This was a radical attempt to get them to pause before stigmatizing a patient, seeing them for the person they are and not reducing them to their age or assumed race, gender, or sexuality. It was equally important to express to them how being blamed for disease transmission complicated sex workers' profession during the early AIDS epidemic. In the underground sex industry, women pride themselves on staying off police radar and avoiding diseases, as their bodies are their labor sources. As the historian LaShawn Harris argues, for many women in the profession, "invisibility was key to their economic survival."[8]

For Lockett, CAL-PEP, and many other industry women, police often worsened matters by harassing women and sometimes increasing their chances for disease transmission. Some officers, for instance, would confiscate condoms in women's possession as "evidence of intent to commit prostitution, or even merely list possession of condoms on the arrest record."[9] In other cases, police would poke holes in condoms found in women's possession and tell them, "Happy hunting." On other occasions, women would be arrested and released after store hours, making it hard to purchase contraceptives. "If they need to continue working to make up for lost time," says Lockett, "they then have to work without condoms, which is dangerous, especially with AIDS."[10] Having students hear and read Lockett's and other industry women's testimonies complicates misconceptions of women being disease carriers and recklessly infecting their clients, and it also illuminates the extent to which sex workers went to prevent HIV/AIDS.

Many health professionals did not necessarily make matters better. In 1985, some doctors purported that "AIDS was going to spread to the heterosexual population through prostitutes," leading CAL-PEP associate and AIDS specialist Constance Wofsy to start Project AWARE (Association for Women's AIDS Research and Education) to combat such unscientific claims.[11] Sex workers especially looked suspiciously at epidemiologists, who were popularly known for "researching people, doing their studies and then walking off and leaving [industry women] holding basically nothing." According to Lockett, "People who sat down and gave some powerful [information], don't have a thing to show for it."[12] Upon entering clinical settings with a positive status and drugs in their bloodstream, some women encountered what anthropologist

Michele Tracy Berger calls "narratives of injustice," which followed three themes: In the first, there was a "lack of information, misinformation, or no information" provided by health professionals to seropositive women. Women reported that hospital staff failed to recommend follow-up visits after informing women of their status. The second theme involved a focus on death, whereby women essentially were told "to go home and prepare to die." The third theme was an intersectional display of judgment by physicians, nurses, and hospital staff who ridiculed women for their ethnicity, drug use, and profession. "When they went in for an HIV/AIDS test and when information surfaced that the respondents were potential substance users," says Berger, "comments were made that resulted in them feeling stigmatized."[13]

It is vital and ethical work to change my syllabus to incorporate such perspectives. Yet how these decisions are responded to, undermined, and dismissed is an ongoing problem in the field. For example, during the first week of instruction, one student complained that no articles assigned were written by physicians. This student was then surprised to learn that one of the authors we read, Vanessa Northington Gamble, who wrote the foreword to the 2021 edited volume *Medicine and Healing in the Age of Slavery*, is in fact a historian and physician. In that piece, Gamble applauds the volume's editors, Sean Morey Smith and Christopher D. E. Willoughby, for highlighting "not just medicine but also other healing traditions." This conceptualization, says Gamble, "dethrones Western biomedical theories and the practices of white antebellum physicians as the exclusive and/or primary locus of study to allow for a richer and more encompassing historiography."[14] Despite learning that Gamble is a physician and historian, some physicians enrolled in the course still took offense to the reading material and its critiques of medicine. Their hostility all but reaffirms the need not only to center perspectives of people from marginalized backgrounds but also to consider the ethics of the positionality of the instructor as well. It raises the question why they felt misled coming into my class and what their expectations were. As many authors in this volume note, we need to go beyond the three Ds in attempts to insert the humanities in our field. That is critical to understanding the positionality and perspectives not only of Black professors but also of women, nonbinary folks, and disability scholars who challenge students to think critically and outside their comfort zone.

What made them feel entitled to question their instructor? Who is authorized to use certain sources and write about specific topics? Does one need to be medically trained to teach the history of medicine from a social perspective? The historians Susan M. Reverby and David Rosner have repeatedly addressed these questions. In their article "'Beyond the Great Doctors' Revisited," Reverby and Rosner note, "Despite assumptions that the racial science and medicine of the nineteenth and early twentieth centuries had faded away, the search for a biological basis for race continues."[15] Some physicians enrolled in the class proved this point, often citing studies in which they participated that resulted in different health outcomes for Black and white subjects. Although they did not explicitly claim innate biological differences, subtle references to there being "something to explain" for disparate health outcomes, while dismissing the impact racism has on one's health, implicitly lean on the side of biological determinism. Conversations stalled, however, when they were asked why few, if any, studies were conducted that compared poor white Americans in Appalachia and upper-middle-class white Americans in Boston or those from other wealthy groups and environments. Rather than having honest conversations about the ways in which race is what the legal scholar Dorothy Roberts calls "a political category that has been disguised as a biological one," the students continued to engage in dialogue that was hostile and opinionated, with minimal critical thinking and examination.[16]

This was a learning lesson that showed me the need for antiracist pedagogy in history of medicine and health care courses that will, in one way or another, branch into medicine, science, and health care. My colleagues Ayah Nuriddin, Elise A. Mitchell, and I had this in mind at the start of the coronavirus pandemic. After the 2020 state-sanctioned murders of Breonna Taylor and George Floyd, to name a few, we felt compelled to put together a public-facing syllabus addressing anti-Black racism in medicine. Recognizing the country's historical record of health crises that have adversely affected African Americans, we believed the emerging (and ongoing) coronavirus pandemic would be no different. Thus, we brought to the public's attention that COVID-19's impact "elucidated [the United States'] long disparate treatment of Black people and centuries-long neglect of Black health concerns."[17] The syllabus consists of sixteen weekly topics, from early medical and scientific theories of racial differences to anti-Blackness in the context of the then-novel coronavirus. By showing a history of anti-Black racism in health and medical education, including enslavement and medical experimentation, the syllabus further illustrates what knowledge is gained and produced through racist practices. It is only by inverting course curricula on histories of medicine and health care so that they start with patients and minoritized subjects who were victims of medical thought and institutions that these goals can be obtained. History is an uncomfortable subject, and a history of medicine conducted outside the three-Ds approach provides alternative ways of learning how the field continues to expand, but also how groups from all ethnicities contributed to biomedical knowledge production. Amplifying their voices, including their skepticism regarding the medical community, shows that no field is perfect, and progress is subjective.

As the bioethicist Annette Dula stated in 1992, medicine and health care "cannot be exclusively biomedical or even ethical." Both fields, Dula continues, must "deal with beliefs, values, and cultural traditions, and the economic, political, and social order."[18] There are few ways that will better prepare us to confront these issues than privileging those who have been historically neglected and their experiences, as Lockett and CAL-PEP did in the 1980s and early 1990s. Their approach required being honest about ways in which HIV/AIDS affected women in the sex work industry differently across ethnicities. Many white women, says Lockett, operated as call girls who were picked up and dropped off at locations, leading to fewer interactions with law enforcement and reducing their chances of street-level sexual violence. Many Asian women operated in massage parlors that were heavily secure, and clients were required to use protection. Poor and working-class Black and Latinx women and trans folks, on the other hand, were also the ones CAL-PEP saw using drugs frequently. This helped the organization structure its outreach. According to Lockett,

> Some of the addicted prostitutes live from day to day. They'll pay their room for one day, and then they have to go out and turn a date before they can get another room. And so, for instance, if it's 2:00 and they haven't had a date, and a guy comes along and says, "Here, I want to date you, but you can't use a condom," they're not going to use a condom. And they've told us that over and over again. So what CAL-PEP tried to do is work with them by telling them the games that they can

play, what they can do to protect themselves and still get the money. We also try to make them understand what the risks are, because some of them really don't know or haven't paid enough attention. . . . That's why we've been concentrating on addicted prostitutes because the career prostitutes don't need us to tell them much.[19]

Leading history of medicine and health care classes from an antiracist pedagogy is an ethical stance. The conservative history of the field means we will be met with resistance for years to come. Rather than discourage us, that should motivate us to keep improving the field and producing good scholars. According to historian Keith A. Wailoo, "Professional and human crisis has spawned the search for meaning and introspection about life, illness, recovery, human suffering, the care of the body, and death."[20] Patients' perspectives should be central in how those stories are told. Their voices need to be amplified, which will force us to read against the grain and question archives that have done the opposite. While medicine continues to advance, minoritized communities—Black communities in particular—continue to suffer disproportionately. Centering marginalized communities—or decolonizing the field—requires us as instructors and learners to see how certain populations are adversely affected during times of crises. Interdisciplinary approaches such as those in medical humanities, disability studies, and Black studies will only improve our field. That includes in the classrooms and in our scholarship.

NOTES

1. James H. Jones, *Bad Blood: The Tuskegee Syphilis Experiment* (New York: Free Press, 1982); Vanessa Northington Gamble, "Under the Shadow of Tuskegee: African Americans and Health Care," *American Journal of Public Health* 87, no. 11 (November 1997): 1773-1778; Susan M. Reverby, *Examining Tuskegee: The Infamous Syphilis Study and Its Legacy* (Chapel Hill: University of North Carolina Press, 2013).

2. Zoe Christen Jones, "Fauci Urges Black Community to Be Confident in COVID-19 Vaccine: 'The Time Is Now to Put Skepticism Aside,'" CBS News, December 8, 2020, https://www.cbsnews.com/news/fauci -black-community-covid-19-vaccine/.

3. Jones.

4. Marshawn Lynch Beastmode Productions, "Sitting Down with Dr. Anthony Fauci," YouTube video, 32:02, April 16, 2021, at 9:00–16:00, https://youtu.be/ODHkJihI2Js.

5. Vanessa Williams, "Among Black People, AIDS Is Taking a Heavier Toll," *Philadelphia Inquirer*, March 11, 1986. Quote from p. 14A.

6. Keith A. Wailoo, "Patients Are Humans Too: The Emergence of Medical Humanities," *Daedalus* 151, no. 3 (Summer 2022): 200.

7. Samantha Majic, *Sex Work Politics: From Protest to Service Provision* (Philadelphia: University of Pennsylvania Press, 2014), 4.

8. LaShawn Harris, *Sex Workers, Psychics, and Numbers Runners: Black Women in New York City's Underground Economy* (Urbana: University of Illinois Press, 2016), 12.

9. Priscilla Alexander, "Role of Prostitutes in Prevention of HIV Transmission: Education + Empowerment = AIDS Prevention," June 1989, Box 1, Folder 72, Women's AIDS Network (WAN) Records, MSS 95-04, Archives and Special Collections, University of California, San Francisco.

10. Gloria Lockett, "Destroying Condoms," in *Sex Work: Writings by Women in the Sex Industry*, ed. Frédérique Delacoste and Priscilla Alexander (San Francisco: Cleis, 1987), 158.

11. Constance B. Wofsy, "Women and the Acquired Immunodeficiency Syndrome: An Interview," *Western Journal of Medicine* 149, no. 6 (December 1988): 688.

12. Gloria Lockett, interview by Nancy E. Stoller, February 14, 1993, Box 5, Folder 35, MSS 2000-6, Nancy Stoller Papers, Archives and Special Collections, University of California, San Francisco.

13. Michele Tracy Berger, *Workable Sisterhood: The Political Journey of Stigmatized Women with HIV/AIDS* (Princeton, NJ: Princeton University Press, 2004), 87–90.

14. Vanessa Northington Gamble, foreword to *Medicine and Healing in the Age of Slavery*, ed. Sean Morey Smith and Christopher D. E. Willoughby (Baton Rouge: Louisiana State University Press, 2021), viii.

15. Susan M. Reverby and David Rosner, "'Beyond the Great Doctors' Revisited: A Generation of the 'New' Social History of Medicine," in *Locating Medical History: The Stories and Their Meanings*, ed. Frank Huisman and John Harley Warner (Baltimore: Johns Hopkins University Press, 2004), 184.

16. Dorothy Roberts, *Fatal Invention: How Science, Politics, and Big Business Re-create Race in the Twenty-First Century* (New York: New Press, 2011), 4.

17. Antoine S. Johnson, Elise A. Mitchell, and Ayah Nuriddin, "Syllabus: A History of Anti-Black Racism in Medicine," *Black Perspectives* (blog), August 12, 2020, https://www.aaihs.org/syllabus-a-history-of-anti -black-racism-in-medicine/.

18. Annette Dula, "Toward an African-American Perspective on Bioethics," *Journal of Health Care for the Poor and Underserved* 2, no. 2 (Fall 1991): 261.

19. Lockett, interview by Nancy E. Stoller, Nancy Stoller Papers.

20. Wailoo, "Patients Are Humans Too," 200.

CHAPTER TWENTY-FOUR

Images and Primary Sources

BEATRIZ PICHEL

As a historian of photography teaching on topics such as medicine, war, and colonialism, I am used to working with sensitive sources with students.[1] History is not pretty, and photographers make sure to show it in all its ugliness. I believe we need to look at and critically examine these documents. Yet doing it in a safe way in the classroom is complicated. What even is a "sensitive" image? Is sensitivity just equivalent to graphicness? And for whom are these images sensitive? In this chapter I disentangle ethical dilemmas that I encounter teaching history with visual sources, as well as the reactions I get from students. My takeaway after years of teaching and research is that visual sources, and photographs specifically, are excellent pedagogic sources to think ethically and critically about the historian's relationship to the past. To achieve this aim, two things need to happen in the classroom. First, instructors and students need to explore photographs and other visual materials as sources on their own rather than as illustrations. Second, students need to be part of the discussion about the ethics of taking, displaying, looking at, and scrutinizing these images.

To put these two principles in practice, I follow three main strategies. The first one is the use of trigger warnings. As the next section will show, using trigger warnings with sensitive images is not always a straightforward decision, and it requires careful consideration of what you want to achieve by including them. My second strategy is to situate medical photographs in their visual, medical, and photographic histories. While I acknowledge that not all instructors in the history of medicine will have training in visual methods, it is important to always look at the contexts of production and consumption if we want to treat images as historical sources on their own. Finally, I explore the reactions I get from students when we debate whether we should look at images of pain and suffering. As Michaela Clark convincingly shows in this volume, there are as many reasons to use medical images as there are not to use them. Discussing these reasons, students soon figure out that the key to this debate is not whether we should look at difficult images but *how.* Under which circumstances, or in which contexts, could looking at sensitive images in the history of medicine be an ethical stance—and on the contrary, which ways of looking and consuming would we deem unethical? It is in this constant dialogue that both students and instructors can critically reflect on the ethical responsibilities of the historian in relation to the people in the past.

Trigger Warnings

It is common practice to use trigger warnings with sensitive images. Trigger warnings, also known as content warnings or content notes, are generally seen as good practice, as they allow students to prepare themselves for what is coming. For instance, I always use trigger warnings when I teach about the photographs of prisoners at Abu Ghraib taken by the American army. The images are so explicit, and the acts committed in the pictures so deplorable, that it is a relatively straightforward decision to use trigger warnings to alert students that the images include torture and sexual abuse. This is particularly important now, twenty years after the photographs first went public. While most of the images are available online, many of my students have not heard of Abu Ghraib or have not seen the photographs before, so I take extra care when I present them for the first time.

Yet it is not always easy to implement trigger warnings in an effective, ethical way because images are always open to interpretation. Graphicness is not the only defining factor of sensitivity, as photographs of seemingly banal subjects can also trigger viewers in different ways, as I examine later in relation to figure A. Viewers actively participate in the creation of meaning; therefore, it is difficult to predict how images are going to be received. Furthermore, as Jessica Carniel notes, there is frequently a "conflation of trauma and offence" in the use of trigger warnings.[2] According to her, while an act of offense attacks individual sensitivities, trauma creates a more "deep-seated" psychological hurt.[3] Not all graphic, explicit, or sensitive images are necessarily triggering; they might just be offensive. Navigating the boundaries between the potentially offensive and the potentially traumatic involves a fair amount of work from the instructor and, I argue, a transparency policy on the choice of images.

An example of these blurry boundaries between the traumatic and the offensive is the use of public health posters related to AIDS in teaching. Following Tasleem J. Padamsee's article on US HIV policy, I asked students to examine public health posters that represented different approaches, from celibacy to activist campaigns based on explicit sex education.[4] But I was fully aware of the irony that, while discussing the benefits of communicating safe sex openly, I was resistant to print out and give students the posters that depicted sexual acts or showed genitals explicitly. I could have used trigger warnings, but what would they have warned of? Was it the nudity, the sex, or both? In this case, it would be wrong to warn about these images because their (mostly gay) sexual content can cause offense—the very warning that these images can be offensive could be offensive toward queer students. But can explicit depictions of sexual acts, for instance sex in a public bathroom, trigger traumatic responses? In this case, instead of using trigger warnings, I avoided posters with visible genitals or explicit sexual acts in case they were triggering. I also explained my decision in class and shared an online link for students who wanted to do more research on their own.

Carniel goes even further in her problematization of trigger warnings, arguing that they can prevent students from engaging with materials. Following scholarship on the pedagogy of discomfort, she argues that using the language of triggers can predispose students to approach the material from a place of trauma and negativity, disempowering them. Based on experience, I would argue the opposite: the use of trigger warnings can make students reflect more deeply about their ethical responsibility in relation to the subjects depicted in the photographs, especially within the history of medicine, health, and disease.

Trigger warnings can help us to create a safe space in which we can discuss conditions that have been historically stigmatized, regardless of whether the photographs are graphic or morbid. The photographic collection of George Edward Shuttleworth is a good example. Most of the images could pass for bourgeois portraits if it were not for the scribblings on the margins. This is the case of figure A (trigger warning:

offensive language for disabilities), which shows two cardboard-mounted portraits of small children, one standing up next to a chair and the other one sitting on an armchair. Both children are well dressed and posing, which shows awareness of being photographed. The content of the images is not the problem. The writing on the margins, however, make these photographs potentially triggering. At the top of the card, a big "Idiots, sporadic cretin" label identifies the disabilities of these children with language that has long been in disuse. The offensive language might put off educators from using these images in class, but the use of trigger warnings could be helpful. A warning similar to the one I used to introduce figure A would allow students who find this language traumatic to prepare themselves, and it would inform the rest of the students that the following content is problematic for particular reasons.

The strategic use of trigger warnings can lead to a productive conversation on ableist language, the ableist gaze, and ableism in the

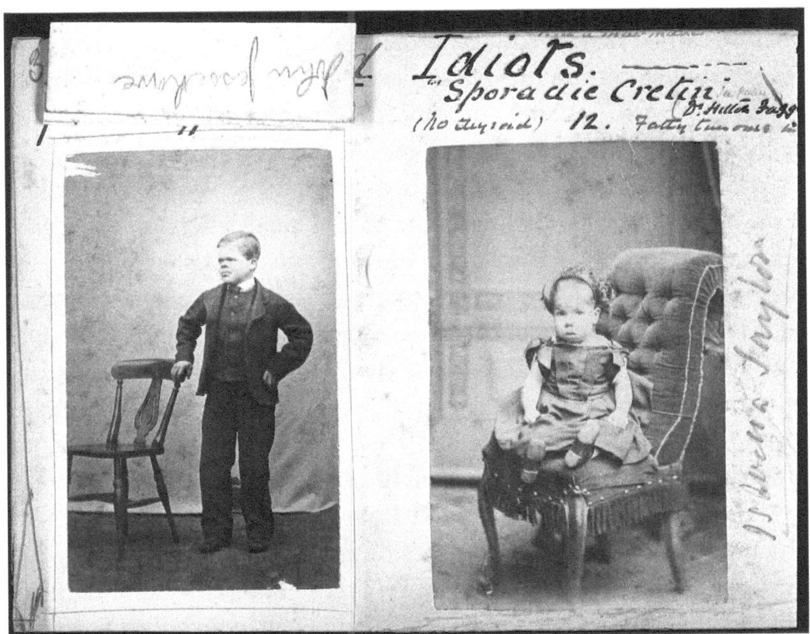

Two photographs of congenital hypothyroidism in a boy and a girl (figure A). Ref. 39110i, Shuttleworth, G. E. (George Edward), 1842–1928 collection. Wellcome Collection, https://wellcomecollection.org/works/c6w4adjb /images?id=wrgepam3. Public domain.

history of medicine. It can also help students to reflect on how they relate to people in the past, highlighting their ethical responsibilities as historians. If we do not refer to people as "idiots" or "cretins" now, should we reproduce images such as figure A uncritically, just because they belong to the past? Stopping the flow of the class, making a pause, and signaling that the content that follows is potentially triggering not only protects students with trauma. These actions also disrupt the idea that the past is ready for consumption, creating pedagogic possibilities to rethink the ethical stakes of historical research in the context of medicine and health.

Photographs as Sources

Trigger warnings can be helpful, but there are other strategies instructors can follow to use visual sources in the classroom in an ethical manner. In my opinion, the most effective intervention we can make is treating images as historical sources rather than illustrations.[5] Placing an image in a PowerPoint only to make materials more visually appealing can be handy, but also problematic. This does not mean that we cannot use certain images in class. Rather, it means that we need to do extra work to use them ethically, and that extra work starts by acknowledging the photographic and medical practices at play.

My classes on the history of medical photography always include three elements: visual analysis of the images, analysis of how the photographs were taken and looked at in the past, and discussion of the medical context that made those photographs possible. I understand many educators in the history of medicine might not have training in visual history or visual methods. I am not saying my approach is the only approach or that you cannot use medical images without deep knowledge about these three aspects. Rather, I mention these three elements as points of reference that help to open the discussion to ethical questions.

For instance, I would never illustrate a class on the history of psychiatry with figure A without spending some time analyzing these photographs. It is not ethical to present these images without examining the implicit ableism of the medical project in which they were created. Yet figure A also contains the potential to rewrite the history of psychiatry from a disability studies point of view. As mentioned before, the content of the images is not problematic per se: they are just children, posing like many middle-class children in the nineteenth century. What would happen if we focused on the children themselves and the stories they tell? This is only possible if we treat these photographs as sources on their own that were created and looked at under specific photographic, medical, and social conditions.

In my teaching, I often use the photographs of female patients diagnosed with hysteria in the late nineteenth century at the Salpêtrière hospital in Paris. Some good examples are the photographs of women with anorexia taken by the photographer Albert Londe for Dr. Wallet and published in *Nouvelle iconographie de la Salpêtrière* in 1892.[6] The very nature of these images is problematic. Wallet's article described anorexia as a disorder characterized by visible emaciation that occurred mainly among young women and that was caused by a woman's decision not to eat. As Erin O'Connor argues, this interpretation only cared about the disease as far as it challenged traditional ideas of femininity.[7] In fact, Wallet's article commented on the lack of menstruation and the size of the breasts in relation to the rest of the body. The photographs reinforced the rhetoric that the "restoration of female health" was the "restoration of gender," as O'Connor states.[8]

The photographs, therefore, bore witness to the emaciation and eventual recovery of weight, providing essential evidence to Wallet's case. Figure B (trigger warning: eating disorders) shows two front-and-back photographs of a white woman, naked from the waist up. The woman is seated on a chair, with a white sheet covering her lower body (a shawl is also visible), against a dark gray background. The use of light and shadows highlights the shoulder blades, spinal cord, collar bone, breasts, and ribs. The photographs follow the photographic protocol used by Londe, who had built a studio especially adapted to the medical research and care at the Salpêtrière. Londe photographed all patients on a stage against a background, which was dark gray if they were naked or light gray if they were dressed, to focus the viewer's attention on the patient's body. Photographs of fully naked men, women, and children were common and routinely reproduced in journals and books. *Nouvelle iconographie* also standardized the photographs by reproducing them all at the same size. These operations helped doctors and readers to compare the images with each other, reinforcing their role as visual evidence.

There is an unusual element in these images that immediately grabs the viewer's attention: the white lines that cover the woman's face. The

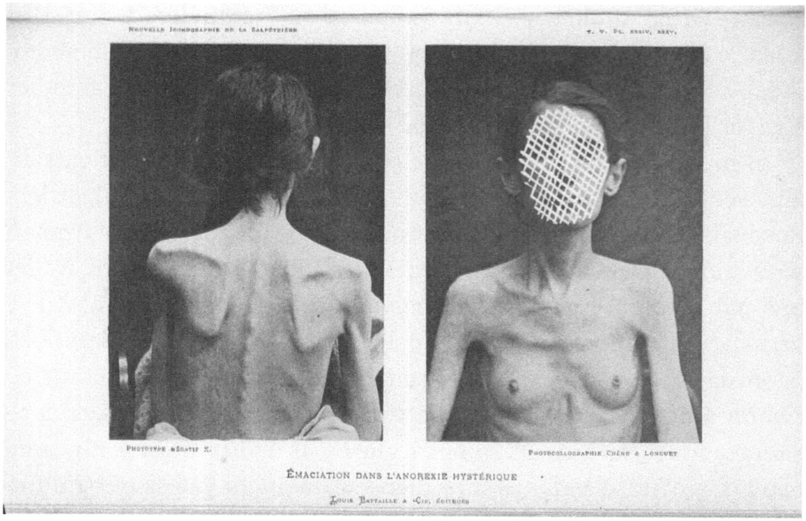

Two photographs of a woman with anorexia (figure B).
From Dr. Wallet, "Deux cas d'anorexie hysterique," *Nouvelle iconographie de la Salpêtrière* 5 (1892). Wellcome Collection, https://wellcomecollection .org/works/hf8jxzxb/images?id=cjhshn6y. Public domain.

article itself does not offer any explanations, although other photographs in the same piece used similar lines to cover another patient's eyes. Looking at other photographs produced at the Salpêtrière or published in the same journal, however, reveals that this was not common practice. While there is nothing remarkable about the way in which Londe photographed this patient, *Nouvelle iconographie* hid the face of the patient at a later stage. Why? We do not know, but drawing attention to the mystery of these white lines allows students to think not only about why these photographs are marked but also why other photographs are not. The fact that the white marks hide the face or the eyes, not the breasts, indicates that it was not the nudity that was problematic but the identity of the patient. Then why are other patients' faces visible?

Engaging with the medical and photographic context of figure B, rather than using it as a mere illustration, raises ethical questions in relation to the role of photography in consolidating problematic medical theories, as well as the representation of patients. Londe's protocol and his masterful use of light and shadows helped to present the illness as a bodily disorder primarily characterized by emaciation and the loss of sexual characteristics. The white lines raise other ethical questions about not just medical practices at the time but also how we, as historians, use medical records. Discussing why this patient had her face defaced can lead us to consider what to do with the thousands of photographs of patients with visible faces. If current medical photography guidelines are based on anonymity and consent in order to protect patients, why are historical patients not offered the same safeguarding? This question goes to the heart of the ethical dilemmas in medical photography. Asking it in class might not lead to a definitive answer, but it will help to establish the parameters of using this kind of material in class. More importantly, it will demonstrate that historical sources are not mere papers in the archive but traces of people's lives. As Adria L. Imada argues in this volume, people in the past are "far more than names and 'data,'" and as historians we need to treat their identities with care. This approach can help students to think about their ethical responsibilities toward people in the past.

Discussing Ethics with Students

But how do students react to sensitive images? Is this even a problem for them? In my experience, once we start paying attention to the ethics of images, these questions become more and more prominent. They suddenly come up in other sessions, proving that students are more than willing to engage with ethics. In fact, at the end of the year, when I ask students to share three things they have learned in my class, many will mention the ethics of photography.

In one of my classes, we always have a debate on whether we should look at photographs of pain and suffering. In preparation for the debate, students need to decide which position they will take and read accordingly.[9] I usually frame these positions as the two ends of the spectrum: "No, we should not look at images of pain and suffering," and "Yes, we should look at these images." The day of the debate, I always start the class by asking who will be on each team, but students rarely identify with one side or the other (some say they will play devil's advocate for the sake of having a debate). The fact that most students do not completely align with either position is exactly what I expect. My role in the debate is to moderate, managing speaking turns and, sometimes, asking further questions based on their discussion. Students quickly realize that the real question in the debate is not *whether* we should look at images of pain but *how*. Under which conditions would it be ethically appropriate to engage with patient photographs such as figures A and B? What would constitute unethical uses of these images? Asking these questions lets us move from the looking/not looking dilemma toward a more nuanced and reflective approach where the onus is not on the image itself but on how historians (and the public) relate to it.

In class, this discussion often leads students to argue that nonconsensual images of pain should not be displayed in exhibitions or shared on social media but should remain accessible to researchers in the archives. But just as Michaela Clark concludes in her contribution to this volume, they acknowledge that this position is not ideal. It also involves further questions that go to the heart of our profession. Why are we, as historians, more equipped than the public to deal ethically with sensitive images? This question is much more difficult to answer, and we do not always reach an agreement. Questions about the ethics of images, therefore, are not only about the images. Interrogating what makes certain uses of images ethical or unethical leads students to think about their own responsibilities as historians.

Conclusion

Teaching the history of medicine with visual materials can be challenging, as many of these images were produced under exploitative circumstances or come with offensive language. Yet sensitive images can become productive material to think about our responsibilities as historians toward people in the past. One of my strategies in class is the use of trigger warnings, which protect students from potentially traumatic content and stop the flow of the class, allowing other students to reflect on why this content might be problematic. For this to work, we need to treat visual materials as historical sources on their own. This means avoiding using images as illustrations and instead delving into their visual and medical contexts. Finally, having open and transparent conversations with students about the ethics of these materials can lead to important discussions on their ethical responsibilities as historians.

NOTES

1. As this chapter is based on my experience teaching undergraduate historians at De Montfort University, I would like to thank my Photography and Conflict students who have discussed these issues with me, making me rethink my position and my arguments over the years.

2. Jessica Carniel, "[Insert Image Here]: A Reflection on the Ethics of Imagery in a Critical Pedagogy for the Humanities," *Pedagogy, Culture and Society* 26, no. 1 (2018): 143.

3. Carniel, 143.

4. Tasleem J. Padamsee, "The Politics of Prevention: Lessons from the Neglected History of US HIV/AIDS Policy," *Journal of Health Politics, Policy and Law* 42, no. 1 (2017): 73–122.

5. On photography as a historical source, see Elizabeth Edwards, *Photographs and the Practice of History* (London: Bloomsbury, 2022).

6. Dr. Wallet, "Deux cas d'anorexie hystérique," *Nouvelle iconographie de la Salpêtrière* 5 (1892): 276–280.

7. Erin O'Connor, "Pictures of Health: Medical Photography and the Emergence of Anorexia Nervosa," *Journal of the History of Sexuality* 5, no. 4 (1995): 535–572.

8. O'Connor, 543.

9. Susan Crane, "Choosing Not to Look," *History and Theory* 47, no. 3 (2008): 309–330; Frank Moeller, "The Looking/Not Looking Dilemma," *Review of International Studies* 35, no. 4 (2009): 781–794.

CHAPTER TWENTY-FIVE

Discomfort and Compassion in the Classroom

CORNELIA LAMBERT

Problems

Practicing the history of medicine requires the contemplation of human suffering. This includes all manner of bodily harm, ranging from the "icky" (blood and guts, fleams and scarificators) to the intangible (nausea, mental illness, pain). The diseased and injured body is part and parcel of history of medicine courses. But a second category, social and political suffering, includes inequality, oppression, and loss or violation of bodily autonomy or identity. This is the stuff of violence; it makes an appearance in every kind of history classroom.[1]

In her book *Staring: How We Look* (2009), bioethicist Rosemarie Garland-Thomson describes the stare as an "intense visual engagement" in which is created a "circuit of communication and meaning-making."[2] When we ask students to contemplate difficult topics, whether in written or visual form, we are asking them to stare. The work of scholars who unravel the act of looking has helped me come to terms with what it means to ask students to "stare" at difficult things. In what follows, I will share how I navigate this rocky ethical terrain. Most crucial have been reflection on my own whiteness and a serious personal goal to hold students' attention in a way that decenters myself and gestures toward decolonization of the classroom.[3]

I have always been drawn to demanding subject matter, and upon reflection I can point to a few reasons.[4] First, I wanted to be a professor so I could talk about serious things, so I could profess what is true and right and necessary to know. Showing difficult-to-process images also appealed to me because my own research, centered in late eighteenth- and early nineteenth-century Britain, concerns looking, in particular, at how members of social classes crafted opportunities to surveil one another. I am especially drawn to instances of transgressive looking—moments that strain what I call "ocular propriety," or the appropriateness of the stare.[5]

Even if unintentionally, I brought this topic to the front of the classroom with me when I became a teacher. After all, what better way to secure student attention than to reveal something they had never experienced? I liked the feeling of being a little transgressive, of showing and talking about things other professors might not. In many ways, talking about the serious stuff was, earlier in my career, about me.

I justified these practices with the argument that digesting difficult truths is part of students' moral and intellectual development. Isn't tension necessary? Socrates challenged his students with questions that had no answers; Martin Luther King Jr. referenced this pedagogically necessary "tension in the mind" in his "Letter from Birmingham Jail." In this way, teaching is the art of directing and narrating attention toward something a student cannot yet do or comprehend, toward what Lev Vygotsky called the "zone of proximal development."[6] This is inherently uncomfortable. Wouldn't exposure to the harsh realities of history help my students become empathic, prosocial creatures? Spending most of my time submerged in eighteenth-century theory, I read again and again Adam Smith's contention, in his *Theory of Moral Sentiments* (1759), that sensing the suffering of others is often the first step in developing sympathy. When we witness the suffering of another, he wrote, "by the imagination we place ourselves in his situation, we conceive ourselves enduring all the same torments, we enter as it were into his body and become in some measure him."[7] In reading about or looking at the suffering of those in the past, my students would be able to imagine their way into a type of communion with others, a link Smith called sympathy. Wasn't it a good idea to help students develop compassion? Wouldn't this make them into better people?

About the time I began teaching, however, changes were afoot that caused me to question aspects of my practice. American culture at large and pedagogical practice in particular began to take the mental comfort of students seriously. I was committed to making my classroom accessible, warm, and fair, but how did that actually play out, with my affinity for difficult topics? Students revealed to me their own histories of abuse and harm and their need for content warnings to protect themselves from further injury. I vowed to honor those requests and to craft content cues into all of my courses.

I was also challenged at this time to interrogate my own whiteness and how it affected my choices in the classroom. I recognized, because history was teaching me, that practices of looking and surveillance are vestiges of colonialist practice.[8] Was my penchant for displaying difficult images related to my whiteness? Was I reproducing past harm when showing pictures of violent acts? And finally, how did I know for sure that what I did in the classroom resulted in a student's developing empathy? Did violent imagery mean anything to students who grew up surrounded by it? What if they were aroused by it? Or prompted to be violent themselves? What if they developed sympathy for perpetrators? And plus, was this empathy business even my job? Just as Saidiya Hartman has asked us to probe how and why we write histories of violence, I needed to ask myself why I taught so often about it, and how to do it ethically.[9] As I became overwhelmed by these questions, the discomfort was, for once, all mine.

Potential Solutions

Conceiving of student engagement with challenging course material as a kind of "stare" has helped me develop practices that facilitate meaningful encounters. Like prolonged visual engagement, a student's interaction with discomfiting material will include both moments of wanting to look and moments of wanting to look away.[10] The stare is a gaze—male, medical, colonial—that rationalizes, names, and in other ways treats the bodies of "others" as "specimen and spectacle."[11] Yet one also insists on a "right to stare," the privilege, Nicholas Mirzoeff argues, echoing Jacques Rancière, that "confronts the police who say to us, 'Move on there's nothing to see here.'"[12] Similarly, classroom engagement with episodes and descriptions of violence is a way for students to rationalize, name, and learn about the past, while also bearing witness to bodily and symbolic harms. If the stare is the basis of our meaning-making in the classroom, it is crucial that we carefully craft how that stare is going to happen and what we are going to ask the student to do with it. Often this means asking them to look, even if it is uncomfortable, even if it is for longer than social mores allow.

My goal is for students to engage with violent and "icky" material long enough to begin to construct new knowledge, to push through discomfort into a space some call "productive discomfort."[13] I still, to a basic extent, believe Smith to be right about the sympathetic effects of comprehending the suffering of others, but I am more aware that I have to construct the right conditions for those effects to occur.[14] In particular, I recommend that professors of health humanities and the history of medicine think critically about and carefully choreograph their actions at three distinct moments: while planning the course, during the first week of class, and at the moment at which students are asked to contemplate or look at the most difficult subject matter.

Planning: Putting Difficult Material in Its Place

It is imperative to think with care about topics and assignments that may be emotionally challenging to your students when designing your course curriculum. This involves being honest with yourself about the volume of difficult material you put in front of students, the placement of that material within the semester-long dialogue, and, finally, the function of that material in achieving your goals for the course. As I order concepts and texts for my courses, I plot the cognitive skills tested by various assignments, counting the confrontation of problematic topics and imagery among those challenges. Except in certain cases, perhaps upper-level coursework, tough subject matter should only be incorporated after the first month or so of the term, when you are assured of your students' ability to treat it with maturity. I also avoid placing difficult material during times of the semester at which students will be distracted: homecoming, midterms, or the week before spring break. Although it adds another hurdle to the daunting task of ordering and organizing a semester, it is worth it to put your most difficult material at a part of the semester in which you can get students' full attention. I also plan out the contours of a semester-long, metacognitive narrative. In other words, I plan out how I will talk to students about the difficult material and what exposure to that material might feel like for them. What tone do I take? How will I disarm them and prepare them for honest engagement? This also helps me know the pace at which to lay the groundwork for a culture of trust and candor in the classroom.

Finally, consider what difficult material will best achieve not only your goal for the students but perhaps *their* goal in taking the course.

The first time I taught Science and the Holocaust, I took a pretty straightforward approach to the material, including some footage of the liberation of Dachau. Of course I warned the students that the material would be difficult (I talk more about this warning shortly); and I waited until I was far enough in the term to know that the students could handle the imagery with maturity. But on the last day of the class, my students, who had all paid very close attention to every part of the course, still ruminated on "why people could be so horrible." While my curricular choices had, indeed, taught them about the Holocaust, I realized that people come to a class like that with a variety of questions they want answered, and that my choice of emphasis had distracted us from what they most wanted to understand.

The second time I taught the course, I dialed back the history of science and medicine readings and replaced some of the imagery I had shared the first time with a unit on the aesthetics of Nazism: the Autobahn, Hugo Boss, and Cathedrals of Light. I also assigned these students a questionnaire at both the beginning and the end of the course, one question of which concerned whether they felt they understood *why* the Holocaust happened. Although this was in no way a scientific or definitive survey, and while I had not queried the first group as I did the second, my questionnaire revealed that this cohort developed a greater understanding of the German public's descent into fascism and support for genocide. In retrospect it seems obvious: what attracted people to Nazism was the beauty it promised them. Juxtaposition of that promise with the ugly reality of genocide helped answer what the students most wanted answered. This epiphany, about crafting the course toward students' particular questions, has influenced my curriculum design ever since.

The First Week of Class: Starting the Metacognitive Narrative

The first week of class is the earliest opportunity to build a foundation of trust and openness with students; it is when to start talking to them about the course as an affective experience, something they may not have ever considered. As the following examples demonstrate, I believe that students should be made aware from the very beginning of the semester that there is tough subject matter ahead, that I am aware that it will be uncomfortable for us, but that there is room for us to talk about what we are looking at and for us to take care of ourselves. This typically means a note on the first or second page of the syllabus detailing course material. For example, a recent course on "the body during the Renaissance and Reformation" included the following *nota bene*:

> This course concerns the history of the human body and we will be discussing—and looking at images of—bodies in all their manifestations. Topics include, but are not limited to, human genitalia, circumcision, sexual maturity, erection, menstruation, sex, transgender bodies, pregnancy and birth, lactation, human "deformity" and difference, death and the dead body, and autopsy. There will be funny moments and things we can laugh about, but our discussions will be completed in a professional manner when serious subjects necessitate it. If you are nervous about these topics, please see me; I am happy to talk.

Both times I taught Science and the Holocaust I drew up a short waiver-like document to share with students on the first day of the course. I asked the students to sign the sheets to signal their full understanding, although I asked neither to see those signatures nor to collect the documents. Instead, I asked them to keep the sheet at the front of their notebooks to remind them of the following:

> We're going to talk about upsetting issues in this course. We're going to look at hard-to-look-at images, talk about death, murder, and unpleasant aspects of the human body and human nature. It is okay if you take a break. It is okay if you cry. I might cry. By signing this sheet, you signal your awareness of what we're about to undertake and a promise to honor your own feelings and the feelings of those around you as we take this journey together.

In each of these cases, I (a) mentioned the subject matter when there was still time to drop the course, (b) gave permission for stress-relieving behaviors like removal from the room and expression of emotion, and (c) set out one example of a potential behavioral outcome indicative of overwhelming emotions. They can give in to both desire to look and desire to look away.[15] These statements communicate to students that I may not know how the material will affect them, but I predict that it will.

How is this different from using trigger warnings? Although practices vary (see, for example, Pichel's contribution to this volume), trigger and content warnings tend to be given immediately before delivery of the offending content, not well in advance. And this would make a certain kind of sense, except that students need time to prepare themselves and manage their feelings. At least one study has indicated that this practice ramps up anxiety rather than providing students with ways to manage their apprehension in healthy and effective ways.[16] This is borne out in anecdotal evidence: a colleague recounts that the only semester a student fainted during the showing of a video re-creating early modern vivisection was, she says, the first semester she warned students at the beginning of class that the imagery might be difficult for them.[17]

Giving early warning allows students to plan. In several instances, I have had students tell me after these announcements that there are particular subjects (usually sexual violence) for which they will need to leave the room. Knowing this, I make a note to myself that I must give a heads-up the class period before I expect to cover the material in class. That way, students can plan to be absent; those absences are excused without question. In some cases, students plan to attend but warn me they will step out. I encourage them to do what they need.[18] After all, Smith warned that, for some people, being faced with the suffering of others can result in visceral reaction and horror. Some people, Smith said, report "that in looking on the sores and ulcers which are exposed by beggars in the streets, they are apt to feel an itching or uneasy sensation in the correspondent part of their own bodies. The horror which they conceive at the misery of those wretches affects that particular part in themselves more than any other; because that horror arises from conceiving what they themselves would suffer."[19] He made no mention of people whose horror arises because they have, themselves, been victims of suffering. To have ignored this reality for so long is one of my biggest regrets about my early years of teaching.

At the Moment

At the moment difficult material is shared, it is important that students have the tools to understand that they are engaged in a stare. Thus, I say so. "Today," I might start, "we're looking at some difficult imagery [reading a challenging passage, etc.]. I don't share this lightly; I understand that it feels weird to look at this, but it is important to our engagement with [topic]." By narrating the reality of the moment, I help students understand that I am not offering this information wantonly, but for a purpose. In their own experiments with the "pedagogy of discomfort," Oda-Kange Midtvåge Diallo and Nico Miskow Friborg model how to be "transparent about [their] own difficulties, doubts, mistakes and learning processes" as instructors.[20] Similarly, I provide a brief description of my own deliberations so that students understand the necessity of the material to the course.

Elsewhere in this volume, Beatriz Pichel advises us to teach any source, whether textual or visual, as material in its own right, not as a method of illustration. In this vein, a tried-and-true technique for the classroom is to ask students to describe the creation of a text or image so that they will understand it in light of the power dynamics that either made it possible or threatened its very existence. Engage their imaginations, as Smith would have us do. In this context, the imagination is a powerful tool for helping students orient themselves beyond the object of the piece, toward its subject.

Do not hesitate to direct students' attention. For example, when teaching about Reconstruction and the Jim Crow South, I have used a variety of techniques to talk about and portray lynching. But the last time I covered the topic, I did as Emily Dickinson instructs—I told the truth but told it "slant." When sharing images of lynched men and women, I blurred the lynched body and focused on the smiling, joyful faces of the lynchers.[21] I realized the irony at the time—erasing the Black body from the picture in order to focus on the white ones? *Was that how I was going to do this?* But the lesson worked, perhaps because the students were not distracted by death. The message about extralegal violence, about white supremacy, was clear. I did offer to show the unedited photograph after class to anyone who wanted to see it. No one did.

The techniques and attitudes shared in this last section concern ways for students not merely to witness but to "bear witness," a distinction made by philosopher Kelly Oliver. To merely recognize the suffering of others is part and parcel of the oppressive structures in which suffering takes place, Oliver argues. My pedagogic goal is what Oliver calls "the pathos beyond recognition," taking the student through discomfort toward bearing witness to "what cannot be seen."[22] I take this to mean the greatest import, the deeper meaning of an image, passage, or event, and what the artifact tells us about the human experience. In these words I am reminded of one (white) American history student who thanked me at the end of the course for talking earnestly about lynching. "Because," she said, she had "never really understood what lynching *was*." I remember distinctly her emphasis on "was," because it communicated to me her understanding of the historical and moral weight of the practice. This kind of reflection is the best for which we can hope: an honest engagement with the full meaning of human suffering, one earned without personal trauma or regret.

NOTES

1. The following friends and colleagues have discussed this topic with me and aided with development of the ideas herein: Kathleen Crowther, Thomas Foth, Aparna Nair, Michael Rotjan, and Rebecca Schmerer.

2. Rosemarie Garland-Thomson, *Staring: How We Look* (New York: Oxford University Press, 2009), 3.

3. Can the university *be* decolonized? Can white scholars fully decolonize their practices? See Edwin Mayorga, Lekey Leidecker, and Daniel Orr de Gutierrez, "Burn It Down: The Incommensurability of the University and Decolonization," *Journal of Critical Thought and Praxis* 8, no. 1 (2019): 87–106; and Karen Buenavista Hanna, "Pedagogies in the Flesh: Building an Anti-racist Decolonized Classroom," *Frontiers* 40, no. 1 (2019): 229–244.

4. Sara Ahmed asks, "What are the implications for feminism that our points of entry are often sore points?" Similarly, why are difficult topics so often my academic destinations? Sara Ahmed, "Feminist Hurt/Feminism Hurts," *Feministkilljoys* (blog), July 21, 2014, https://feministkilljoys.com/2014/07/21/feminist-hurtfeminism-hurts/.

5. Cornelia Lambert, "Barefoot Children in a 'Fine Room': Robert Owen, Adam Smith, and Social Regeneration in Scotland," in *The Routledge History of Poverty, c.1450–1800*, ed. David Hitchcock and Julia McClure (London: Routledge, 2020), 349.

6. L. S. Vygotsky, *Mind in Society: The Development of Higher Psychological Processes*, ed. Michael Cole et al. (Cambridge, MA: Harvard University Press, 1980), 84.

7. Adam Smith, *The Theory of Moral Sentiments* (1759; Amherst, NY: Prometheus Books, 2000), 3.

8. A full citation list for this topic includes Frantz Fanon, *Black Skin, White Masks* (New York: Grove, 1952); and Edward Said, *Orientalism* (New York: Pantheon, 1978). A more recent scholar on whose work I depend is Nicholas Mirzoeff, *White Sight: Visual Politics and Practices of Whiteness* (Cambridge, MA: MIT Press, 2023).

9. Saidiya Hartman, "Venus in Two Acts," *Small Axe* 12, no. 1 (June 2008): 1–14.

10. Garland-Thomson, *Staring*, 79.

11. Nora L. Jones, "Embodied Ethics: From the Body as Specimen and Spectacle to the Body as Patient," in *A Companion to the Anthropology of the Body and Embodiment*, ed. Frances E. Mascia-Lees (Oxford: Blackwell, 2011), 73.

12. Nicholas Mirzoeff, *The Right to Look: A Counterhistory of Visuality* (Durham, NC: Duke University Press, 2011), 1. For more on the police officer's assertion that there is "nothing to see," see Jacques Rancière, Davide Panagia, and Rachel Bowlby, "Ten Theses on Politics," *Theory and Event* 5, no. 3 (2001), https://doi.org/10.1353/tae.2001.0028.

13. The origins of this term are murky and, with a few exceptions, it is more common in business discourse than in texts about education. See, for example, Alexandra Barron, "Productive Discomfort in the Classroom: Teaching *Boys Don't Cry*," *New Review of Film and Television Studies* 3, no. 1 (2006): 75–84.

14. My optimism is tempered by my colleague and friend Aparna Nair's statement to me about students and empathy that "if they don't have it by the time they get to us, it is not something we can remedy." Personal conversation.

15. Garland-Thomson, *Staring*, 79.

16. See Benjamin W. Bellet, Payton J. Jones, and Richard J. McNally, "Trigger Warning: Empirical Evidence Ahead," *Journal of Behavior Therapy and Experimental Psychiatry* 61 (December 2018): 134–141.

17. Kathleen Crowther, personal conversation.

18. We have to be prepared, however, for when students opt in, then regret it. See also Barron Lerner's essay in this volume.

19. Smith, *Theory of Moral Sentiments*, 4.

20. Oda-Kange Midtvåge Diallo and Nico Miskow Friborg, "Subverting the White Cis Gaze: Toward a Pedagogy of Discomfort, Accountability and Care in the Anthropology Classroom," *Teaching Anthropology* 10, no. 4 (2021): 27.

21. In his 2005 exhibit "Erased Lynching," artist Ken Gonzales-Day digitally erased lynching victims from commemorative photographs in order to achieve these same ends. Ken Gonzales-Day, "Erased Lynchings," Ken Gonzales-Day's website, accessed November 22, 2024, https://kengonzalesday.com/projects/erased-lynchings/.

22. Kelly Oliver, "Witnessing and Testimony," *parallax* 10, no. 1 (2004): 79, 80.

CHAPTER TWENTY-SIX

The Problems with Partners

SHARRONA PEARL

Partnership is a funny kind of word. It can mean a lot of different things. An initial association is likely of equity: two or more people or entities sharing an ongoing set of obligations, transactions, or benefits in a fair and (ideally) predetermined fashion. There are romantic partners, and business partners, and pedagogical partners, and, of course and inevitably, corporate partners, which is usually capitalist-speak for sponsors. And, of course and inevitably, so many of these partnerships are not equal at all.

That is not always a bad thing. When it comes to collaboration on a university course, there is deep value in both recognizing one's own needs and requirements and making space for those of others. Sometimes that is an ethical process, the result of careful negotiation, ongoing education, and thoughtful evaluation. Sometimes it is a coerced or nonconsensual situation of voluntoldism or professional obligation. And sometimes it is in the awkward space in between, where, due to personal or professional reasons, one has no choice but to accede to one's partner in ways that are not exactly coercive but not entirely comfortable. I developed an undergraduate course that partners with the Mütter Museum in Philadelphia, whose anatomical and pathological specimens as well as wax models were originally collected for medical teaching purposes. This is an ethically fraught space that has historically leaned into a model of freakery, displaying and marketing human remains (many of which are without provenance) in often sensationalist and indeed exploitative ways. To its credit, the museum is currently undergoing a major (and overdue) ethics audit.[1] In partnering with it, I had to consider my own responsibilities and ethical obligations around collaboration and critique as both a guest of the space and the instructor within it. I share these concerns with the students at the outset, inviting them to consider whether there can be ethical partnerships or relationships in such settings, and what role we can play as guests and visitors who are not experts in museums but still hope to participate in the process of change.[2] The Mütter's history is one that historically centers white male doctors, and the museum has engaged in problematic acquisition, branding, and display practices that perpetuate dehumanizing conventions for nonwhite and nonelite people as well as perpetuating harmful messages on disability and difference. The history cannot be changed. Yet the future branding, display, and acquisition practices must, as must the way we teach in and about these spaces.[3]

Partnering with museums can be weird. Partnering with museums as part of a course that explicitly critiques their specific practices, displays, collections, and histories can be really, really weird: there is the ethical consideration of working with a problematic entity (which most are, so it becomes a question of degree, pedagogical upsides, and possibilities for interrogating that very question); on the other hand, it may be a different kind of ethical breach to work with a partner in order to critique them. There is also the question of going intentionally into a space that may be painful or difficult for students, making it imperative to be absolutely clear in all course materials about what the partner institution is and what it does.[4] In the case of the Mütter Museum, we have been very lucky that it too is in a moment of self-critique, allowing us to partner with it in a mutually agreed-on process of intervention. This self-critique is really valuable and important from both a pedagogical and, indeed, a practical point of view. Our partnership created space for a productive discussion and modeled for students in real time the long, difficult, and slow process of bringing about change while still insisting on interrogating both the ethics of the space and the ethics of partnership with problematic institutions. Such partnerships are delicate and require a lot of advance preparation and negotiation, as well as very clear guidelines and expectations for both the institution and those participating in (and conducting) the course. They can also be incredibly rewarding. Students are sometimes in the best position to make lasting change, or at least to communicate the need for it. Even as we investigated the ethics of the Mütter itself, we considered the challenges of intervening in complex institutional structures that are dependent on revenue and embedded in traditions and expectations. Such constraints, common to many institutions, do not always allow them to immediately adhere to what some believe to be the best ethical practices in collection and display.

In consonance with broader trends in the museum world, the Mütter is undergoing a process of reflection and reevaluation. It is actively auditing how its objects were acquired, what can be put on display, and, importantly, how to navigate all of the items in its collection that were neither consensually donated nor displayed with respect. Our partnership with the Mütter provided very real value in analyzing these questions in a grounded setting, offering students the opportunity to be a part of the reevaluation process. This raised an ethical problem and, equally, an institutional one: How can I balance my university's and students' needs and expectations with those of this institutional partner while at the same time offering a unique hands-on pedagogical experience with valuable assignments and rich reading and discussion?[5]

In this essay, I first discuss the practical and ethical considerations in building a course in partnership with an institution with a problematic past and present, as well as how to navigate this structurally and pedagogically. I highlight practical strategies to create an effective and ethical partnership. I then offer a comprehensive guide on how to develop and manage assignments and assessment in an ethically fraught context. Such a course, by design, hopes to intervene in the unethical and indeed racist acquisition and display practices of the museum partner that has generously invited the students in to curate potential exhibits. Students develop valuable curation and exhibition skills that consider questions of space for objects and notes, while at the same time working with museum professionals to learn the fundamentals of archival research, storytelling, and museum pedagogy. So how should we proceed in creating, developing, and continuing such partnerships in a pedagogical framework, if we should at all? To borrow from capitalist-speak: it is all in the ask.

The Ideal and the Possible

Let us discuss what is needed to build a museum-embedded course on the ethics of museums and bodies on display. The syllabus for such a course likely engages with a lot of the readings referenced in this very volume. It tracks the history of museum practice and collections with an emphasis on the highly problematic nature of both the acquisitions and how they are shown. This in and of itself is a valuable pedagogical intervention that would likely appeal to a wide range of students. But let us say you want to ground this theory in practice, perhaps as part of a partnership with a local institution.[6] And let us say that institution, for reasons of its own (and these are, of course, relevant and need to be considered, as they will absolutely frame what is possible within this collaboration), is eager to work with a local institution of higher learning. You have your own particular commitments and intellectual obligations, which include as the first priority critically engaging with museum practice broadly and, indeed, with this museum in particular. That is the very point of the course. How, then, to ethically proceed in a partnership that has as its starting assumption critique of that very institution? And also, is working with an unethical institution itself unethical, affirming through partnership the problematic practices with which it engages?

That is where things get exciting. And tricky. And diplomatic. Let me be clear: many museum professionals, like those at the Mütter, are highly eager to engage in precisely this kind of critique and discourse. Many institutions are well aware of the challenges they face going forward as they consider the legacies of their past (and, let us be honest, present) practices. And many are going to be wary of inviting in a group of people whose explicit goal is to do exactly that. So, again, it is all in the ask. It is all in the framing. It is all in the way the endpoint is presented, and that should include something that benefits everyone.

We have now defined both the problem—a course that, *by design*, critiques the very institution in which it is embedded and with which it is, perhaps problematically, partnering—and the ideal outcome: a process and final project in which the students work with the institution and its staff to meaningfully intervene in the museum's process, practice, or (and) planning. That is a lot to hope for, particularly given resource limitations, institutional constraints, short term lengths, and (often) lack of background on the part of students. And even if that precise model is not possible, there is rich scope for a powerful experience with rewarding and meaningful assignments. But let us think big, and let us think concrete. How should such an ideal course be structured, and what are the models of collaboration and partnership that would support it?

Models and MOUs (Oh My)

The first step is to have a series of conversations with your institutional partner. What would they like to see come out of this collaboration both in the short term as a final project and in the long term as an ongoing relationship? Are they open to providing space for student work (with due caveats for what the outcomes may be)? Will they support such work with resources including vitrines or cases, in-house printing, staff time, and access to the collections? Will they link to an online exhibit and provide their branding for it? What about language—do they use the same terms to refer to their collections that you might want your students to employ? How often will students visit the museum, and how much access will they have to the holdings with which they will be working? There could be one visit, or three, or indeed every class may be held in the partner institution. Given the limits of time and student experience, perhaps the museum could isolate a series of objects or questions for the students to both frame the experience and ensure the projects are doable and within the scope of the museum's interests. Remember that partnerships work two ways. The course is offering student labor and feedback, but it is asking for a considerable amount of labor and commitment from the museum. This may be in line with the museum's mission, and the institution may or may not be visitor supported. It is worth investigating whether your home institution offers resources, including admission to the museum, funding to support staff time, internet space for a potential online exhibit, and possibly marketing and publicity for the outcomes. The answers to these questions will help frame the assignments and expectations for the course.

If these initial conversations are successful, the museum might be open to a longer discussion about critique and possible interventions. It may already be engaged in a process of reflection on unethical exhibits, marketing approaches, or historical collection practices. Students may participate in these investigations in their own work, exploring a particular question in the context of the museum's own history. This may be embedded in the larger context of how other institutions have dealt with, say, unethically acquired exhibits, problematic language in displays, gift shop items that are not respectful, and other historical legacies that may generate income and still need to be interrogated.[7] Again, this may not be a viable framework for the collaborative aspect of the course; it may still be an option for a final assignment outside the partnership with the institution. The key is to lay out expectations very carefully and perhaps even with an MOU (memorandum of understanding) that may need to be vetted by the university's general counsel. This should include photo releases and clear guidelines as to who may use any outcomes from the course, such as social media posts and summaries of the projects and indeed the final projects themselves. Be very careful to emphasize and instantiate the partnership relationship; swooping in as saviors would not only undermine the very ethic of collaboration, ignore community-centered practices, and overpromise outcomes, it would be sure to doom the relationship to failure.

Assignments and Evaluation

There are a lot of possible approaches to assignments, final projects, and evaluations in a museum-embedded course. The syllabus will likely comprise a range historical, theoretical, ethical, and anthropological readings. These will sit alongside introductory material on basic museum storytelling, generating big ideas and coherent narratives, and best practices in curation and presentation.[8] As the assignments will be consonant with the overall goals of the course, the balance between curation and critical analysis will reflect this approach. The institutional partner may play a role in determining whether projects will be displayed or used, but the evaluation and grading are of course always up to the instructor. (Let me say that again: the final evaluation is *always* up to the instructor. This is also something that must be made clear from the outset.)

This is a humanities course. There will be writing. There *should* be writing, even as it is also potentially an exhibition or curation course. And the writing makes most sense as part of the critical and theoretical material on the syllabus. Students can be asked to write regular response papers that connect course readings to the specific partner institution or one or more of the objects with which they are working. These papers can be structured to connect to the applied work the students are doing either directly or in a more theoretical way, which can then be brought together during class discussion. While the final project will be a more hands-on, practical, and grounded piece of work, students can be asked to write a theoretical analysis of their contribution. This will connect more broadly to the critical and humanistic elements of the material. I like to include a close reading assignment early in the term to guide students through a deep examination of a particular source; they may or may not use that source in the final project. They certainly will use these skills.

There is going to be group work. There is no way around it, and given the partnership model that brings students together with an institution and its staff to develop grounded techniques in museum practice, it is a valuable reflection of how stuff actually gets done. Also, time is short, and the work is hard. While group projects make some things more difficult, they also allow for greater depth and nuance through the (sometimes) equitable division of labor and pooling of creative, academic, and practical resources and skills. There is huge value in collaboration. That does not mean that everyone will like it.

Students tend to resist group projects for a number of (very understandable) reasons, including the difficulty of scheduling meeting times outside class, the "freeloader effect" from students not pulling their weight, and the overall challenges of collaboration and compromise. These are valid concerns—and also, too bad. It is the instructor's responsibility to structure assignments in ways that limit these challenges, including by providing ample class time to work on projects and divide up responsibilities in ways that limit the necessity of meeting outside class; assigning a self- and group evaluation for each student to complete individually alongside the final project to allow them to express the relative contributions of each member; and scheduling regular check-ins and steps to the assignment to make sure that everyone is on track and working well together.

The final project may be an exhibit, the parameters of which will be determined in part by the institution's interests and constraints. It may also be a theoretical critical analysis of some aspect of the museum's own practice. The museum may suggest areas for examination, or you or the students may decide to explore a particular issue. These may include provenance, donation protocol, access to collections, use of objects in marketing and publicity, relationship to donors, collection philosophy, or museum language and branding. As I hope is clear, my philosophy as a scholar starts with generosity and the value of collaboration and discussion; while this kind of course necessarily and deliberately includes critique, framing the partnership with a narrative of "fixing" or "exposing" is likely neither generative nor productive. The fact that the partner institution is willing to work with your course is a sign of good faith and interest in thinking with your students; if the institution is open to suggesting areas of examination, this is another mark of trust and engagement.

In either case, students should be guided through the stages of their work, starting with the generation of big ideas and big-picture approaches to their specific collection of objects or concerns. An initial assignment, early in the term, may ask the group to gather and create a (running) annotated bibliography with a list of questions and possible stories they wish to tell. This will lead to a project proposal and presentation that will set the groups up for success and provide a nice point to make sure that everyone is participating and working well together. I like to include a peer critique and feedback element that allows students to practice giving generous and generative feedback while taking advantage of collective wisdom and individual student strengths. If possible, students should also be in contact with museum staff to coordinate practical details regarding text size and color, space constraints, and overall branding to maximize the possibility that their work will be used. The course instructor should be involved in these consultations both to support student efforts and to negotiate any challenges that might arise from unexpected philosophical differences.

Following the project proposal assignment, students can be asked to produce a blog or social media post highlighting their works in progress. Such entries help students extract and share key points about their work while also helping them develop the ability to discuss their projects from different perspectives. The institutional partner may be eager to share these either under its own auspices or as linked posts. Students will continue to work with syllabus readings and writing assignments; as with the close reading exercise, these can be tailored to track their work with the museum either directly or indirectly. The final project and accompanying reflective essay will be determined by the nature of the ask: an exhibit will have parameters different from a pamphlet for visitors, which will in turn be different from a policy brief or set of institutional guidelines. At this point, there should be no surprises. (There are always surprises.)

Executive Summary

This is a joke but also not really. In a somewhat programmatic essay, I am going to conclude with a really programmatic synopsis (because maybe someone outside academia will read this? Probably not, but what is the introduction or conclusion if not a kind of executive summary, actually?).

The Problem: The task is to craft a course and a series of assignments that critique the very institution with which the course is partnering in as ethical a fashion as possible while acknowledging the challenges of collaborating with an ethically fraught space.

The Strategy: As the instructor, you should clearly communicate about the institution and your perspective and goals to both the students and the partner, discuss outcomes and projects in advance, and negotiate roles and parameters before assignments are determined.

The Assignments: The students should engage in regular critical writing alongside and in conjunction with museum-related skill-building exercises and presentations. These can include response papers, close reading exercises, annotated bibliographies with key questions highlighted, blog and social media posts, peer critiques, and final projects.

The Evaluation: Final evaluation can be based on a balance between process and product, depending on the goals and structure of the course. If the focus is more historical and research oriented, students should be given an opportunity to demonstrate how that is reflected in the work. If it is a more curation-based approach, the evaluation will have a different emphasis. Museum staff may be invited to share reflections and feedback, but the instructor is always the one giving the grades. (If there are grades. There may not be grades. Even still, everything else stands.)

As programmatic as this is, there is no one-size-fits-all course and no way to anticipate all the challenges in any given pedagogical endeavor, let alone one that brings in additional partners and stakeholders. At the same time, working with partners is exciting, introduces a new environment, and helps students and faculty develop valuable connections while engaging with the wider community and seeing how change happens—or does not—in real time. While there are major considerations in terms of pedagogical responsibilities in entering into relationships with ethically fraught partners and how we help students navigate those spaces, there is also great scope to raise these considerations both specifically and generally in the classroom environment. And while students may or may not be able to participate in the work of the institutional partner, they will certainly learn a great deal about grounded ethics and what it takes to create change. They will have an applied opportunity to consider not only what not to do but—sometimes, and ideally—what to do and how to do (less harm).

NOTES

1. On the Mütter's process, see, for example, Edward Helmore, "'Macabre Curiosities': Top US Medical Museum Confronts Skeletons of Its Past," *Observer*, August 13, 2023, https://www.theguardian.com/us-news/2023/aug/13/medical-mutter-museum-philadelphia-specimen-ethics; Michael Tanenbaum, "Mütter Museum Weighs Ethical Concerns over Its Online Exhibits and YouTube Videos Displaying Human Remains," PhillyVoice, May 12, 2023, https://www.phillyvoice.com/m%C3%BCtter-museum-philly-online-exhibits-human-remains-ethics/; Malcolm Burnley, "What the Hell Is Happening with the Mütter Museum?," *Philadelphia*, September 24, 2023, https://www.phillymag.com/news/2023/09/23/mutter-museum-ethics-controversy/; and Maura Judkis, "A Museum's Historic Human Remains Are Now the Center of an Ethics Clash," *Washington Post*, July 27, 2023, https://www.washingtonpost.com/lifestyle/2023/07/26/mutter-museum-controversy-philadelphia/.

2. There is a wide range of opinions about how the Mütter ought to proceed. For two different perspectives on disability representation, see Riva Lehrer, "Philadelphia's Mütter Museum Is Reviewing Its Collection of Human Remains. Here's Why That Matters for Disability Representation," *ARTnews*, June 16, 2023, https://www.artnews.com/art-in-america/columns/mutter-museum-op-ed-riva-lehrer-disability-1234671870/; and Riva Lehrer, "Where All Bodies Are Exquisite," Opinion, *New York Times*, August 9, 2017, https://www.nytimes.com/2017/08/09/opinion/where-all-bodies-are-exquisite.html (note that Lehrer has long been an advocate of the Mütter). See also Katrina Jirik's essay in this volume.

3. For a discussion of museum ethics with respect to the museum's own artifacts, see Jane Nicholas, "A Debt to the Dead? Ethics, Photography, History, and the Study of Freakery," *Histoire Sociale/Social History* 47, no. 93 (2014): 139–155.

4. Scholars continue to debate both the efficacy and the imperative for trigger warnings; in the case of this course, the syllabus serves to underscore both the contents of the Mütter and the key focuses of the course. Jeannie Suk Gersen, "What If Trigger Warnings Don't Work?," *New Yorker*, September 28, 2021, https://www.newyorker.com/news/our-columnists/what-if-trigger-warnings-dont-work.

5. This course was designed around "community-based learning" pedagogy. Tania D. Mitchell, "Traditional vs. Critical Service-Learning: Engaging the Literature to Differentiate Two Models," *Michigan Journal of Community Service Learning* 14, no. 2 (2008): 50–65.

6. Valerie Farnsworth, "Conceptualizing Identity, Learning and Social Justice in Community-Based Learning," *Teaching and Teacher Education* 26, no. 7 (October 1, 2010): 1481–1489.

7. Katrina Jirik has investigated these issues in the Mütter in this volume.

8. These might include classic exercises like those presented in John Hennigar Shuh, "Teaching Yourself to Teach with Objects," in *The Educational Role of the Museum* (London: Routledge, 1999), 80–91; and Elaine Heumann Gurian, "What Is the Object of This Exercise? A Meandering Exploration of the Many Meanings of Objects in Museums," *Daedalus* 128, no. 3 (1999): 163–183.

CHAPTER TWENTY-SEVEN

Teaching Graduate Students about Ethics

SHANNON K. WITHYCOMBE

"Week 7: Ethics . . . ?"

The words on the paper stared out at me, filling me with fear and dread. It was January 2020 and I was writing the syllabus for my first-ever graduate historical methods course. It was not the first time I faced a blank week while designing a syllabus, nor would it be the last, but this was different. It was not just that this week was blank; it was that I had no idea of even where to begin. I knew that I wanted to talk to my graduate students about ethics but had no idea what that meant.[1]

As a historian of medicine, I am very familiar with the slippery nature of the meaning of ethics. It is a concept that changes over time, encapsulates different skills and ideologies for different groups, and moves power around academic communities, often away from their subjects of research. I was fairly comfortable talking with my undergraduate students about how to determine what *ethical* could mean when looking at medical treatments, scientific studies, or public health experiments in the past. Comfortable, that is, in primarily convincing them of the complexity of ethics.

But practicing good ethics is not something typically taught in a graduate history program, especially in a history of medicine track. I had not been trained in this. The teaching of ethics to history of medicine graduate students became my ethical dilemma. Who was I to decide what good ethics were for academic history? I had spent the last ten years teaching the history of medicine at a midranked state university, and I held no political sway or position of power in an organization like the American Historical Association or the American Association for the History of Medicine. I had no book awards, no prominent students, and no research funding for my second book project. I was to some, by virtue of identifying as a woman, not qualified for the "hard" stuff.[2] What gave me the authority to teach others how to become ethical historians?

Analyzing the Past

I began by looking back to my own graduate experience. I came to graduate school knowing nothing about the history of medicine and nothing about graduate school. I just knew that I liked reading stories of illnesses in the 1800s. It never occurred to me, in the seven years I spent earning my MA and PhD, that power imbalances were responsible for my constant companions in grad school: insecurity, imposter syndrome, and general self-loathing. I just thought I was bad at history and a slow learner.

I sat silently in most graduate seminars, convinced that any book I found confusing or irrelevant was mostly likely one I was not smart enough to understand. No one ever pointed out that my life experience was probably very different from that of Thomas Kuhn, so maybe relating to his "influential" and "essential" book would be a foreign experience.[3] This anxiety only increased when it was time to conduct my own research for a dissertation about pregnancy loss. I read through personal letters of women who described knowing their pregnancy was over by the physical and emotional heaviness they felt, and hospital accounts of patients who had lost multiple children in pregnancy or soon after giving birth. I shoved my sadness down deep, convinced that any emotion I showed would reveal me to be an inadequate researcher. I was pretty sure Kuhn never cried in the archives, and so I was a bad historian if I did.

I left that training with my PhD but still with little understanding of how to ethically think about my fellow historians, my research subjects, or the responsibilities that come with teaching the history of medicine.

Examining the Present

When it became my turn to teach, I had no idea how to teach ethics effectively or ethically, but I knew how I did *not* want to teach. I did not want my students to leave their graduate education feeling ignorant, isolated, and as though they were failures.

My first ideas about teaching ethics came from my undergraduates. I was trained at a large state university in the upper Midwest and can remember a mere handful of discussions about Native Americans, but only as historical subjects and not as scholars. So when I started teaching at a large state university in the Southwest and looked out at Native students sitting in my classroom, every lecture about the "beginning" of America or about "public" health suddenly rang false, hypocritical, and patronizing. In my second year of teaching, while I was giving a lecture on eugenic sterilizations in the early twentieth century in Virginia, a student raised her hand and asked, "What about the sterilization of Navajo women at the Gallup hospital?" This was a hospital 140 miles from where we were sitting, and these were women my student knew. I had no answer. I knew nothing about those women, that practice, and that violence. But I realized that "I didn't learn that in grad school" could not be my excuse. Continuing to neglect the experiences of Native Americans meant that I was perpetuating centuries of harm. I needed to listen and learn from new sources, scholars, and voices, including the young people sitting in my classrooms. I needed to find ways to tell my students that their communities and stories mattered.

But it was not enough to find an article here or there that discussed medical genocide of Indigenous groups in North America; I also wanted to talk with my students about why it was seemingly acceptable for me to go through seven years of graduate training without anyone asking me about colonialism, Native sovereignty, or the role of medicine in genocide. Instead of teaching my undergraduate students about the Columbian Exchange of disease from the classic article by Alfred Crosby, I began instead to teach about why Crosby, in the 1970s, was inspired to transform the narrative of Indigenous depopulation following European arrival on the North American continent into one about differential population disease immunities.[4] And how the virtual absence of Native scholars in the field of history helped to sustain this narrative for the next forty years, downplaying active genocide by colonizers.

My students, living in a state often mistaken for a different country, wanted social justice, and I was committed to helping them. Soon after, we all faced a global pandemic that highlighted long-standing racial and ethnic health disparities, due to centuries of systemic racism and colonization. Historians of health cannot, ethically, insist that good history only exists outside politics like social justice.[5]

I took these questions I was asking myself to my graduate students, asking them to think about power and their own position. I asked, What do you think about using patient records as sources? How would you feel if your stigmatized illness or body became public in this way? In what ways have you been harmed by a medical system, and how does that allow you a unique entry point into understanding a source? I came to realize that good ethics was teaching about the humanity within, and the power deployed by, the history of medicine.

By the time I had published my first book, earned tenure, developed a dozen undergraduate and graduate courses, and presented over thirty conference papers, I was, I suppose, an "established scholar." But I still felt on the losing side when talking with senior scholars and university administrators, as well as when reading the words of the president of the leading historical association bemoaning practices I was proud of.[6] I wanted to go rogue, to tell my students that maybe Kuhn was not a good model for them, to inspire them to speak out against racism in the archive, colonialism in the profession, and why I, as a white, cishet, able-bodied woman, was not the ultimate expert for any of what they needed to learn about the ethics of history. But how could I push my students to question and critique the system that paid for my childcare and enabled me to keep doing work that I loved?

Shaping the Future

Looking back over the readings I have assigned and discussions I have engendered both in graduate seminars and in professionalization workshops I run for our graduate students, one of the points I keep coming back to is that historians are fully formed people shaped by their lives. This may sound simplistic, but it is something I never considered in the first ten years of my academic training.

My professors and mentors asked me how the evidence I found in sources pointed me in certain directions, how the attendees of a conference reacted to my findings, and how my preliminary readings revealed gaps in the literature. But they never asked about my personal development as a historian, my emotional state after a day in the archives looking through sources about loss and death, or how my experiences as a white woman, a former biological researcher, and a person with depression created a filter of seeing (and not seeing) that only I could possess.

I became more and more convinced that part of my job as a history professor was to help my students to see how the creation of history is deeply shaped by the life experience and socialization of the historian. When I began that first methods syllabus for graduate students, I typed "something about positionality" under the first day. I read through Frances Maher and Mary Kay Thompson Terreault, Mitsunori Misawa, and Renée Martin and Dawn Van Gunten and spent most of those readings nodding vigorously and whispering "WTF?" at how new it all sounded to me.[7] Yes, this was where my mind had been going when listening to Indigenous students contradict a narrative I presented in class, but why had I never heard of these works before? Why was positionality a topic of interest in "teacher education" but not part of the training of graduate students who were supposed to become teachers?

Now I weave throughout graduate courses multiple opportunities for my students to consider the positionality of each author we read and to build up an understanding of their own positionality. As I tell them on the first day, positionality is the combination of all the avenues an individual has to power. We start with some of the more traditional avenues considered in academia: training, adviser, and employment. But then I push the students to consider as many pathways to power as they can: race, ethnicity, class, gender, sexuality, ability, family experience with higher education, generational trauma at the hands of academia or medicine, and on and on.

When we read the words of Thomas Holt, we talk about not only how his life experience being socialized as a Black man in America shaped his arguments about race-making and the writing of history but also how his positionality as a Black historian in 1995 at the University of Chicago shaped what he wrote, and perhaps what he did not write.[8] The more I model considering the positionality of our authors, the more the students begin to recognize their own positionality. In our first meeting, I model this for them: "As someone socialized as a white woman who studies anti-Black biology constructed in early twentieth-century infant mortality science, I know that I can never know what racism feels like and cannot speak to that. I need to listen to the Black women whose infants died in premier hospitals in the 1920s and the Black families still facing higher infant mortality rates than most Anglo-white families in the United States." As others, such as Ayah Nuriddin and Antoine Johnson, have brilliantly shown in this volume, by acknowledging the positionality of historians and historical actors and ensuring diversity across all identity spectrums for both, we can begin to redress violence perpetuated by the historical profession.

Teaching at a university with a large Indigenous student population, I sought to learn as much as I could about decolonizing methodologies and history in my first year on campus. Going through that process inspired me to take certain aspects of decolonization to all historical subjects. One of the tenets of decolonization is redressing the generations of extractive research on Indigenous communities by academics. I introduce this to my students through Linda Tuhiwai Smith's work and through the Native American and Indigenous studies call to arms by Alyssa Mt. Pleasant, Caroline Wigginton, and Kelly Wisecup.[9] These scholars highlight the importance of working toward historical scholarship as a mutually beneficial activity between researchers (still largely white) and Indigenous communities. I ask my students, What will your research give back to the people and community you are studying?

But then I extend that question to every person and community we study. We read Saidiya Hartman and Vanessa Holden and think more critically about who is in the documents, the images, and the stories we access in archives, and who remains today from those families and communities.[10] I ask my students to reflect on how their research could bring harm. Just as I do not want them to learn that they are timeless, contextless researchers, I do not want them to be socialized into thinking the people of the past they encounter are isolated cases, or abstract identities, or faceless figures.

But there are so many other people, with their own experiences and positionality, involved in making history, and it is vital for students to understand that as well. I do an exercise with my graduate students to help them imagine all the people and motivations involved in making history. I ask them to think through every step a source must go through for it to become part of a historical production. Who wrote things down and who did not? Why do people write some things down but not others? What people would be involved in keeping the source and passing it along to others? What people would be involved in the decision to donate the item to an archive? What people would be involved in deciding whether the source was appropriate for the archive? What people would be involved in cataloging and organizing the source? What people would enter that archive, look at that source, and see it as valuable? What people would approve the publication of an analysis of that source? By the end we have a chalkboard full of professions and positionalities, a crowded landscape of power and privilege.

Reflecting on this web of power, I ask my students to consider the possibility that sometimes being a good historian means *not* showing off your latest find. Sources are not prizes to be won or treasures to discover. Not every artifact, story, or person from the past is something that needs to be, or should be, exposed, amplified, and analyzed.[11] Being a good historian is not about how much stuff you can find but rather about how responsible and thoughtful you are with what you find. Figuring out whether a source should be publicized, named, or reproduced is not an easy decision and not teachable in a static policy, but I urge my students to think about consequences and possible harm with every bit of research they do from the very beginning.

I still feel the traces of insecurity when I teach ethics. A small part of me expects one of the students to ask, "What is your expertise in this?" I spent years convinced that because all the experts around me were not talking about ethics, this meant they had already agreed on it and expected me to know. But as I get older and more "established," I have come to realize that sometimes, senior scholars do not talk about something because they do not know either. I may not train graduate students to have all the answers about ethics in the history of medicine, but maybe that is not the right goal. Maybe teaching ethics is teaching students to keep asking the question, What is ethical history? As Richard McKay stresses in this volume, ethics is a process, not a destination.

Recently I sat down with a first-year, first-generation student whom I think is brilliant and listened to them describe the microaggressions they faced in graduate courses, ones that were not acknowledged, much less addressed, by the person running the seminar. While I could talk about privilege blindness among academics and how many professors fear conflict in the classroom, I could not provide them with answers beyond, "I wish they were all better." When I later asked why the student felt comfortable coming to me with this issue, they simply said, "Because you see me." This was when I realized that I do have authority to teach ethics; most of us do. It comes from seeing.

Teaching ethics to graduate students is about providing skills for them to see, acknowledge, and deal with power differentials. And it is always easier to see a power differential when you are on the losing side. I often get asked at national conferences, "So are you looking to get out of New Mexico?" Countless historians assume that if I hoped to be a "good" historian, I was in the wrong position. This profession values training at private universities, positions at private universities, and teaching the "smart" (code for white and upper-class) kids. I have lots of experience with looking in from the outside of power and influence. My students, at a Hispanic-serving university in an extremely poor, minority-majority state, have lots of experience with being on the losing side and do not understand what it is like to feel entitled. But by seeing them, by helping them see the power differentials at work every day in their lives, and by convincing them that their voices are valuable in the profession of history, I can teach them ethics.

NOTES

I would like to thank all the amazing students, undergraduate and graduate, I have had the honor to teach at the University of New Mexico. I also never would have begun this path reflecting on how to teach more ethically without Tiffany Florvil, who turned me on to Holt and Hartman and who spent countless hours talking with me about problems in the profession at various Albuquerque playgrounds. Finally, I would like to extend a huge thank-you to Courtney Thompson, Kylie Smith, and all the participants in the working group that gave rise to this volume. My teaching and scholarship will forever be changed for the good.

1. I am beyond excited that students in my future graduate methods courses get to read this entire volume.

2. I was informed in graduate school by a senior scholar that the one female academic who dared to take on Michel Foucault in a major way was, and I quote, "crazy."

3. Being able to converse flawlessly about paradigm shifts was currency in my graduate program and one I often lacked. If you are not familiar with the concept, see Thomas Kuhn, *The Structure of Scientific Revolutions* (Chicago: University of Chicago Press, 1962), or just look him up on Wikipedia.

4. Alfred W. Crosby, "Virgin Soil Epidemics as a Factor in the Aboriginal Depopulation in America," *William and Mary Quarterly* 33, no. 2 (1976): 289–299.

5. Unfortunately, the pandemic did not convince all historians of this. The president of the American Historical Association, James Sweet, published an article in which he argued, "Doing history with integrity requires us to interpret elements of the past not through the optics of the present but within the worlds of our historical actors." James H. Sweet,

"Is History History? Identity Politics and Teleologies of the Present," *Perspectives on History*, August 17, 2022, https://www.historians.org /research-and-publications/perspectives-on-history/september-2022/is -history-history-identity-politics-and-teleologies-of-the-present. See Courtney E. Thompson and Kylie M. Smith's introduction to this volume for more discussion of Sweet's writings.

6. Sweet doubled down on his crusade against presentism and politics in an interview with David Frum that appeared in the *Atlantic.* Sweet not only critiqued "tweets, or at least blogs or essays online," as "radical" but also argued, "Subjugating history to politics has inherent risks." David Frum, "The New History Wars," *Atlantic*, October 3, 2022. See also Barron H. Lerner's and Susan M. Reverby's contributions to this volume for more on the Sweet scandal.

7. Frances A. Maher and Mary Kay Thompson Tetrault, *The Feminist Classroom: An Inside Look at How Professors and Students Are Transforming Higher Education for a Diverse Society* (New York: Basic Books, 1994); Mitsunori Misawa, "Queer Race Pedagogy for Educators in Higher Education: Dealing with Power Dynamics and Positionality of LGBTQ Students of Color," *International Journal of Critical Pedagogy* 3, no. 1 (2010): 26–35; Renée J. Martin and Dawn M. Van Gunten, "Reflected Identities: Applying Positionality and Multicultural Social Reconstructionism in Teacher Education," *Journal of Teacher Education* 53, no. 1 (2002): 44–54.

8. Thomas C. Holt, "Marking: Race, Race-Making, and the Writing of History," *American Historical Review* 100, no. 1 (1995): 1–20.

9. Linda Tuhiwai Smith, *Decolonizing Methodologies: Research and Indigenous Peoples* (London: Zed Books, 1999); Alyssa Mt. Pleasant, Caroline Wigginton, and Kelly Wisecup, "Materials and Methods in Native American and Indigenous Studies: Completing the Turn," *Early American Literature* 53, no. 2 (2018): 407–444.

10. Saidiya Hartman, *Wayward Lives, Beautiful Experiments: Intimate Histories of Riotous Black Girls, Troublesome Women, and Queer Radicals* (New York: Norton, 2019); Vanessa Holden, *Surviving Southampton: African American Women and Resistance in Nat Turner's Community* (Urbana: University of Illinois Press, 2021).

11. I love Richard McKay's personal reflections on this in this volume.

CHAPTER TWENTY-EIGHT

The Ethics of Teaching Future Health Professionals

LAN A. LI

In March 2020, I founded the Medicine, Race, Democracy (MRD) Lab.[1] Conceptualized in collaboration with Ricardo Nuila and Fady Joudah, two accomplished writer-physicians, the MRD Lab focused on engaging with community clinics as sites of social and therapeutic innovation. These sites included acupuncture colleges, churches, mosques, and charitable organizations in Houston that provided health care services beyond the Texas Medical Center. Members of the MRD Lab worked to explore how community-built clinics fostered racially inclusive environments, financially accommodating infrastructures, and creative therapeutic modalities. We studied the complexities of ethical persistence among our partner organizations in sustaining their mission and among ourselves in sustaining the lab.

This final chapter reflects on how the MRD Lab enacted ephemeral critical pedagogy built on critical studies of race, colonialism, and gender. It aligns particularly with Kylie Smith's chapter on the urgent recognition of continuities between the present and the past in reparative historical writing, Richard McKay's chapter on the ethical orientations required for responsibly engaging with the past, Ayah Nuriddin's chapter on reframing narratives of historical violence through "Black power methods," and Antoine Johnson's chapter on antiracist pedagogy as an ethical position. I have been especially aware of critical methodologies as a longtime filmmaker and media producer, where I work to carefully shape anticolonial visions of Asian medicine. These positions further inform my own work as a scholar on the dynamic epistemic and ontological domains of global East Asian medicine. I enact social justice through aesthetics, an ethical praxis that undergirded the MRD Lab.

An aesthetic praxis is aspirational. It requires constant attention to what remains. To this end, the MRD Lab encouraged students to engage with uncertainty. Lab members created videos, exhibitions, poems, essays, maps, and podcasts as modes of civic engagement across multiple semesters. They generated original research on nonuniversalist, open-ended, and reflexive narratives in the health humanities. In 2021, we supported around two dozen prehealth students, six graduate students, and ten external advisers. We hosted eight research clusters where students took on projects that aligned with their interests and trained them in writing literature reviews, designing critical cartographies, engaging in oral histories, and ethnographic fieldwork. They learned how to endure unfamiliar environments, confront differences, and witness each other's transformations. More importantly, they would understand transformation as an iterative process, never complete and always "emergent," as Michael Fischer would say.[2]

Feminist Critical Pedagogy as Self-Preservation

Moral action is complicated. I am not a philosopher, but I understand that intending to do less harm does not always result in less harm. Ethical pedagogy does not operate on a clean axis of good intentions leading to good outcomes.[3] As feminist ethicist Elizabeth Brake has emphasized, the same moral reasoning can produce contradictory results; those acting ethically may still inflict harm.[4] For instance, liberal egalitarian principles can justify aid for disadvantaged groups yet also enable gentrification. Equal-opportunity rhetoric can support community connection while also ignoring local community values. Moral frameworks are context-dependent; they are not inherently good or bad. Like Nana Osei Quarshie's radical retelling of the pharmakon, practitioners and patients both act as "false prophets."[5] They absorb and inflict harm within political institutions of hope.

The logic of critical pedagogy has long yielded contradictory outcomes. It has justified solidarity while imposing inherently exclusionary principles. Critical pedagogy has supported progressive ideologies while reinforcing power hierarchies. To this end, feminist scholars, such as Elizabeth Ellsworth and Kathleen Weiler, have challenged the dogmatic elements of Freirean critical pedagogy by highlighting how the political and philosophical tendencies of critical theory have often assumed universal binaries between the oppressor and the oppressed.[6]

Yet, these binaries seduce and obfuscate. They contribute to fantasies of liberation. They bolster repressive paternalism. They feed into essentialist identity politics. They amplify savior sentimentality.

Feminist critical pedagogy is a different political project. Rather than assuming one universal subaltern group, it recognizes inherent diversity and the potential for conflict within that diversity.[7] Embracing partial perspectives is a political tactic. It intellectually foregrounds incommensurability to nurture competence and cooperation across difference. At the same time, it does not promise to successfully dismantle false universalism. It contributes to emancipatory and diversity discourses with promising liberation. Within this model, students occupy complex positions as both being disadvantaged and privileged, silenced and well-resourced.

We built the lab based on an acute awareness of difference. The lab validated difference by taking an open and queer approach to identity, which my colleague Jia Hui Lee has described in our conversations as a "radical commitment to unpredictability." It was through unpredictability that the students could explore the paradoxes and tensions embedded within their own identities. It allowed them to confront difference and dissonance without acting under ideological promises of total liberation.

My own positionality did not invoke or assume solidarity. I was a new professor at Rice University in Houston, a city that was not yet my own, at an institution wrestling with its own politics and ethics. My subjectivity rested unsteadily at the intersection of geographic, socioeconomic, and institutional vertices. I came from an artist-immigrant family; my financial experiences were shaped by opportunities as a Pell Grant recipient. I did not speak for my students. But I could facilitate wonder and curiosity as an ethical praxis of feminist critical pedagogy.

Cultivating Self-Preservation through Self-Trust

The inception of the lab was serendipitous. It began during the perplexing beginnings of the COVID-19 pandemic in spring 2020 when no one quite knew how to handle the stifling demands of lockdown and the depressing surge of race-based hate crimes. What did it mean to do something? How could we support students beyond articulated aphorisms? I looked to my own research funds and used it to support individual students who had lost their summer internships. It was a modest attempt. I reached out to Eddie Jackson, a former chemistry major and QuestBridge scholar. We initially focused our work on reviewing publications about the role of Asian medical practices in cancer research. But what started as a database-building project organically evolved into a public humanities endeavor in the form of a podcast—a twelve-episode anthology that trained teams of students to research, write, perform, and edit personal and scholarly narratives in the health humanities. Eddie named the series *meta-stasis*, and we started production on the first episode, which outlined Eddie's personal experience with cancer in their family.

Over two years, Eddie gained experience in researching, interviewing, writing, recording, and editing podcast episodes. The team grew, and I applied for additional funding to hire more students. The research project evolved into a research podcast, then a research cluster, ultimately becoming the MRD Lab. The *meta-stasis* podcast modeled the lab's pedagogical mission of providing a platform for student self-realization.[8] It allowed students to contemplate their relationships with themselves, their family histories, and immigration patterns in Houston. We used methods across the health humanities that critically examined these relationships while developing diverse skills in subjects ranging from literature reviews to data visualizations. Students crafted and shared poems, essays, and podcast scripts that allowed them to expand and deepen their engagement with civic life.

As Eddie reflected on their experience in the MRD Lab, they observed that the lab was integral to developing self-confidence and self-trust. They explained,

> I think the most important thing I learned was to trust myself, you know? The MRD Lab was an experience that allowed me to really develop a lot of amazing skills. I never thought in a million years that I would have been podcasting. . . . I got to be a storyteller. Like that was kind of amazing. I'm terrible at telling stories. But I felt like the podcast was a way for me to act more creatively and with more agency. I'm a chemistry major, and nothing in chemistry is really that creative. I'm going to be honest—it is one of the hard sciences! And I felt myself almost lacking, you know? I wasn't feeling fully fulfilled in my academic studies. And I got to really fulfill myself with this.[9]

For Eddie, the hard sciences had left a void. As Rita Chiron, Sayantani DasGupta, and others have described, narrative practice can construct meaning to shape this void.[10] Although Eddie did not actively seek out narrative practice as a mode of self-transformation, the MRD Lab created a community and infrastructure that supported Eddie's transformation through developing concrete skills. The lab fostered mutual care, enabling students to thrive at their own pace according to their individual needs. When some students faced medical emergencies, peers stepped in to help. We implemented a peer-to-peer mentorship system staggered across different cohorts. The lab served as an inclusive space where graduate students collaborated with undergraduates on original research related to their dissertations. For instance, Yesmar Oyarzun, a graduate student in anthropology, worked with Eddie and Juliann Bi as assistants to develop a digital skin cancer exhibition focused on skin of color. Yesmar virtually presented this project, cleverly titled "undertones," at the meeting of the Society for Social Studies of Science.[11]

As the lab's acting director, producer, and project manager, I worked to involve students in research, storytelling, design, and publication. I made sure that all students had shared access to essential software, including the Notion.io page, Descript account, and WiX backend. They were involved in every stage of research and outreach. I drew on my background in art, music, film, and design to train students in developing personal narratives. We worked collaboratively, using the digital humanities as a critical tool for community research and pedagogy.

Community and Intersubjectivity

As the MRD Lab grew, I introduced Eddie to Sophia Peng, a prehealth student majoring in anthropology. Together, Sophia and Eddie developed a podcast series about acupuncture in Houston. Neither had prior experience with acupuncture, but they inspired each other with a shared sense of wonder about different modes of health and healing. Both Eddie and Sophia were creative, reflexive, and committed students. More importantly, they supported and trusted each other. I assigned Eddie and Sophia to shadow interns at the American College of Acupuncture and Oriental Medicine so that they could familiarize themselves with the field of acupuncture and gain insights into how the college navigated diversity, community engagement, therapeutics, and politics.

Following each shadowing session, Eddie and Sophia recorded debriefing notes and transcribed them using Otter.ai. Over five months, with the help of assistant editors Jameson Horton and Lauren Ginn, we interviewed acupuncturists for the podcast series that Sophia's friend titled *Point Break*.[12] The result was a carefully edited podcast covering a range of topics, including fertility, cancer, heart attacks, and irritable bowels. For instance, Eddie and Sophia met an American College of Acupuncture and Oriental Medicine intern named Charlie Tsing, whose previous experience working as a game designer led him to become interested in Asian healing practices. Sophia and Eddie were fascinated to discover that Charlie had even designed his own gamified card deck for memorizing herb names in Pinyin. We purchased decks for the team.

This embedded approach of shadowing interns, transcribing notes, recording, editing, and producing a podcast fundamentally transformed Sophia and Eddie. It provided them with sustained experiential intersubjectivity through a dynamic production process. Sophia reflected on her experience:

> People talk about experiential learning all the time. But being at the acupuncture college felt very immersive. I felt like I understood something that I only conceptualized in my head in a whole new way. . . . One of the most important things I learned was this understanding of an area of clinical practice through seeing it and reflecting on it more so than literature review or traditional academic forms of learning. We didn't have an obligation to present our findings in a particular academic light, which I think was also revolutionary for how we could process it ourselves.[13]

Sophia's engagement was not passive. Working with the production team and shadowing interns allowed for a fully "immersive" intersubjectivity. She was not familiar with acupuncture but appreciated its cultural legacy without understanding its history or politics. But to understand its history and politics, she had to experience its complexity. I only required that throughout the process, Sophia and Eddie ask questions about what they did not know. They had to reflect, wonder out loud, and ask more questions. They had to actively confront the limits of their knowledge and skill. As Sophia described, there was no "obligation" other than to talk, edit, and share. The intersubjectivity facilitated the learning.

Intergenerational Interrogations

While the *Point Break* podcast enabled self-realization through active collaboration, the lab's poetry research cluster, Personal Reflections, fostered self-realization through intergenerational inquiry.[14] I originally created this cluster to support Summer Nguyen, a former student interested in her grandfather's experiences as a surgeon during the Vietnam War. Summer's family history shaped the direction of research cluster that would be focused on the experiences of Southeast Asian communities in Houston. I paired Summer with Brandon Ba, another prehealth student, whose family had relocated from Burma (now Myanmar) to Texas. The scope exceeded my expertise, and I contacted Sam Lê, a local poet knowledgeable about Vietnamese street culture in Houston, to mentor Summer and Brandon in their craft.

With Sam's guidance, Summer and Brandon undertook a series of poetic exercises and experimented with different forms of writing, reflection, and revision. Summer composed a series of poems based on interviews with her grandfather, Nguyễn Tiến Dy. She deepened her understanding of her own family dynamics and observed what she called "narrative holes" in her conversations with her elders. She observed her grandfather's fondness for America despite its role in the Vietnam War that forced his displacement. Summer explained that her work aimed to seek "a new understanding of omission" and to reckon with the generational distance that she experienced within her family. A tremendously talented poet, Summer ended her series with a piece titled "The Orange Tree," one of the first poems she wrote in a café with Brandon. Sam's instructions were to write a poem without abstractions, which Summer composed from her grandfather's perspective:

> I was sitting
> at the orange tree
> when the orange fell.
> It was unripe.
> Pale, premature, small, inanimate
> when it broke its life-sustaining ties.
> Falling brutally,
> the ground beating him,
> Leaving bruises irreparable.

He laid there, fresh juices
bleeding before the rotting began.
All while
I was sitting.[15]

Summer and Brandon collectively wrote over a dozen poems, theorizing their own lives through intergenerational inquiry. They interrogated family members to fill the silences. They also mentored new cohorts who produced essays, poems, paintings, and digital art. Sophia Peng later joined the Personal Reflections cluster and also worked with Sam Lê, who remained a steady guide throughout the process. As the research cluster grew, we workshopped pieces with external mentors, including Outstanding Bean, Houston's poet laureate, and Woods Nash, a poet and ethics professor at the University of Houston. The student poets considered their positionalities and how pursuing health careers forged a narrow sense of obligation and inhibited their desires. They asked what it meant to be part of generations of health professionals—nurses and doctors—within broader conditions of statecraft, empire, and migrant labor. They confronted feelings of resistance and friction that remained unresolved.

Conclusion

bell hooks, Cherríe Moraga, Lorgia García Peña, and others have famously discussed the importance of communing in academic spaces to allow for self-realization, transgression, and expansion.[16] As Peña observed, building a community is antithetical to academic institutions that encourage prehealth students to succeed based on a model of individual success. Peña writes, "It is through engaging with one another and through difference—through theorizing from our flesh, as Latina feminist writer Cherríe Moraga invites us to do—that we become better scholars, better teachers, and wiser human beings."[17] Critical pedagogy that engages with race, capitalism, and gender moreover offers models of ethical practice that is grounded in meaning-making and self-discovery.

The stakes of these forms of inquiry are high. They are intense and immediate. They mimic classrooms outside the traditional classroom setting, resembling seminars beyond the seminary walls. These spaces are bound by practices of care, generosity, and shared power. They demand scholars to engage with ethical issues of expansion, inclusion, and justice. Health humanities labs are spaces of experimentation and possibility.[18] Yet, they are also ephemeral. The research clusters in our lab sit dormant after I moved from Houston to Baltimore. Students graduate, and grant funding expires. Still, the groundwork remains.

The MRD Lab created a space for students to commune, step outside their flesh, and interrogate it.[19] They shared doubt and celebrated accomplishments in a paternalistic education system that reinforces competing model minority stereotypes. Other labs have done the same. The Environmental Data Justice Lab at the University of Toronto is an Indigenous-led space that appropriates colonial research methods.[20] The Humanities Lab at American University facilitates ongoing, experimental, and open-ended initiatives.[21] The Heat Lab at the University of California, Los Angeles, uses thermal inequality to understand carceral medicine and the environment.[22]

There remains a cost to doing less harm. The MRD Lab was built to support students—to help them survive. It was an equitable space where students were paid for their work. It allowed them as much freedom to learn and create in the back and front ends of our digital platforms. Still, the lab isn't without its limits. For instance, it can do better in its commitment to boycotting its website host, WiX, an Israeli company. While we wait for the resources to move the entire website to a different host, we can match the cost of hosting on WiX in the form of donations to Palestinian refugees. As the lab transforms in and out of dormancy, so will its mode of enacting ethical persistence. Aesthetic, ethical praxis again remains aspirational as we calibrate against the conditions of harm by remaining attuned to the deficiencies of good intentions.

NOTES

1. MRD Lab, homepage, accessed November 22, 2024, https://www .mrdlab.org. At its inception, the lab was funded by Rice University's Chao Center for Asian Studies, Rice University's Provost's TMC Collaborator Fund Seed Grant, and Rice University's BRIDGE Systemic Racism and Racial Inequality Grant. Special thanks to all of the students who were a part of the lab: Adarsh Suresh, Alicia Leong, Alyssa Cahoy, Aruni Areti, Aysel Rizvi, Bilal Rehman, Brandon Ba, Danica, Eana Meng, Eddie Jackson, Emily Ma, Emma Cooke, Esther Lee, Jameson Horton, Jason Lee, Julianne Bi, Katherine Wu, Kaylah Patel, Kim Jones, Laura Ginn, Lily Weeks, Linda Wu, Madison Zhao, Mandy Quan, Pearl Zhang, Saagar Dhanjani, Sabariah

Mohamed Hussin, Sally Yan, SJ Zanolini, Sriya Kakarla, Taylor Phillips, Tian-Tian He, Yesmar Oyarzun, Yi, and Yifan Wang. Additional thanks to special advisers and contributing faculty who assisted me in training the students across the research clusters: Cher Vincent, Danyal Rizvi, Gregory Sparkman, Guangming Li, John Paul Liang, John Sparagana, Laura Napier, Liyen Chong, Melissa Bailar, Pierce Salguaro, Rick Mizelle, Sam Lê, Tani Barlow, Thalia Micah, and Victoria Massie.

2. See Michael M. J. Fischer, *Emergent Forms of Life and the Anthropological Voice* (Durham, NC: Duke University Press, 2003).

3. By "ethics," I mean what can be described as applied ethics and practical ethics. Tom Beauchamp has described ethics as "any use of philosophical methods to treat moral problems, practices, and policies in the professions, technology, government, and the like." Tom L. Beauchamp, "The Nature of Applied Ethics," in *A Companion to Applied Ethics*, ed. R. G. Grey and Christopher Heath Wellman (Hoboken, NJ: John Wiley and Sons, 2005), 3.

4. For more on applied ethics and the contradictory results of moral reasoning, see Elizabeth Brake, "Rebuilding after Disaster: Inequality and the Political Importance of Place," *Social Theory and Practice* 45, no. 2 (2019): 179–204.

5. For an excellent example of anti-colonial historical ethnography, see Nana Osei Quarshie, "Archives of False Prophets: Inventing the Future in a West African Psychiatric Hospital," in *Psychiatric Contours: New African Histories of Madness*, ed. Nancy Rose Hunt and Hubertus Büschel (Durham: Duke University Press, 2024), 43–67.

6. To review the early work on the limits of critical pedagogy, see Elizabeth Ellsworth, "Why Doesn't This Feel Empowering? Working through the Repressive Myths of Critical Pedagogy," *Harvard Educational Review* 59, no. 3 (1989): 297–325; and Kathleen Weiler, "Freire and a Feminist Pedagogy of Difference," in *Debates and Issues in Feminist Research and Pedagogy: A Reader*, ed. Janet Holland, Maud Blair, and Sue Sheldon (Clevedon, UK: Multilingual Matters in association with the Open University, 1995), 23–44.

7. For a survey of the growing literature on the limits of critical pedagogy and the rise of feminist critical pedagogy, see Susan Gabel, "Some Conceptual Problems with Critical Pedagogy," *Curriculum Inquiry* 32, no. 2 (2002): 177–201; Ilan Gur-Ze'ev, "Feminist Critical Pedagogy and Critical Theory Today," *Journal of Thought* 40, no. 2 (2005): 55–72; and Joe L. Kincheloe, "Critical Pedagogy in the Twenty-First Century: Evolution for Survival," *Counterpoints* 422 (2012): 147–183.

8. *Meta-stasis* (podcast), homepage, accessed November 22, 2024, https://www.metastasispodcast.com.

9. Eddie Jackson, quoted in "The Medicine Race Democracy Lab: A Tour of Community Clinics and Collaborative Self-Transformation," in *Innovation in Arts and Health* (Pressbooks, forthcoming).

10. These include, but are not limited to Trisha Greenhalgh, "Narrative Based Medicine: Narrative Based Medicine in an Evidence Based World," *BMJ: British Medical Journal* 318, no. 7179 (1999): 323–325; Rita Charon, "Narrative Medicine: Attention, Representation, Affiliation," *Narrative* 13, no. 3 (2005): 261–270; Maura Spiegel and Rita Charon, "Editing and Interdisciplinarity: Literature, Medicine, and Narrative Medicine," *Profession* (2009): 132–137; Ronald Schleifer, "Narrative Knowledge, Phronesis, and Paradigm-Based Medicine," *Narrative* 20, no. 1 (2012): 64–86; Sayantani DasGupta, "Narrative Medicine, Narrative Humility: Listening to the Streams of Stories," *Creative Nonfiction*, no. 52 (2014): 6–7.

11. To access the full "undertones" exhibition, see Yesmar Oyarzun, Juliann Bi, and Eddie Jackson, "undertones," *meta-stasis* (podcast), accessed November 22, 2024, https://www.metastasispodcast.com/undertones.

12. To access the full season of *Point Break*, see Sophia Peng and Eddie Jackson, *Point Break!* (podcast), MRD Lab, accessed November 22, 2024, https://www.mrdlab.org/point-break.

13. Sophia Peng, quoted in "Medicine Race Democracy Lab."

14. To access the full content of Personal Reflections, see "Personal Reflections," MRD Lab, accessed November 22, 2024, https://www.mrdlab.org/reflections.

15. Summer Nguyen, "The Orange Tree: A Military Doctor's Choice," MRD Lab, 2021–2022, https://www.mrdlab.org/orangetree.

16. On radical feminist literature about communing in academic spaces to allow for self-realization, transgression, and expansion, see bell hooks, *Teaching to Transgress* (New York: Routledge, 1994); Cherríe Moraga and Gloria Anzaldúa, eds., *This Bridge Called My Back: Writings by Radical Women of Color*, 4th ed. (Albany: State University of New York Press, 2015); Lorgia García Peña, *Community as Rebellion: A Syllabus for Surviving Academia as a Woman of Color* (Chicago: Haymarket Books, 2022); María del Carmen Salazar, "A Humanizing Pedagogy: Reinventing the Principles and Practice of Education as a Journey toward Liberation," *Review of Research in Education* 37 (2013): 121–148.

17. Peña, *Community as Rebellion*, 43.

18. For a recent reader on health humanities pedagogy, see Craig M. Klugman and Erin Gentry Lamb, *Research Methods in Health Humanities* (Oxford: Oxford University Press, 2019).

19. On competition, visibility, and invisibility, see Cathy Park Hong, *Minor Feelings: An Asian American Reckoning* (New York: One World, 2020).

20. To learn more about the Environmental Data Justice Lab and its efforts examining the relationships between data, pollution, and colonialism in Canada's Chemical Valley through community-based and Indigenous-led research, see "EDJ Lab," Technoscience Research Unit, University of Toronto, accessed November 22, 2024, https://technoscienceunit.org/people/lab/.

21. To learn more about the Humanities Lab and its aims to promote interdisciplinary and collaborative humanities projects, see https://www .american.edu/cas/humanities-lab/.

22. To learn more about the UCLA Heat Lab and its research in interdisciplinary thermal inequality (focusing on the relationship between heat, health, and the built environment), see "Heat Lab," UCLA Institute for Science and Genetics, accessed November 22, 2024, https://socgen .ucla.edu/research/heat-lab/.

Conclusion

The Expanding Ethics of Doing Health History:
A Personal Journey

SUSAN M. REVERBY

Plagiarize
Plagiarize
Let no one else's work evade your eyes
Remember why the good Lord made your eyes
So don't shade your eyes
But plagiarize, plagiarize, plagiarize
Only be sure always to call it please "research."
—*Tom Lehrer, "Lobachevsky," 1951*[1]

If we don't read the past through the prism of contemporary social justice issues—race, gender, sexuality, nationalism, capitalism—are we doing history that matters? This new history often ignores the values and mores of people in their own times, as well as change over time, neutralizing the expertise that separates historians from those in other disciplines.
—*James H. Sweet, American Historical Association president, "Is History History?," 2022*[2]

Before the 1990s, historians often kept questions of professional ethics at the back of their minds, but seldom on the tip of their tongues.
—*Antoon De Baets, Responsible History, 2008*[3]

Ethical principles for the complicated concerns that come up when you do history have a history of course, but I was never taught to consider the difficult ones. Nor, alas, have many of my younger colleagues, or this volume would not be necessary. My concluding essay, then, bookends Richard McKay's opening piece in that it demonstrates a historian's ethical journey, but from a different time and space. The vague historical ethics that prevailed as I entered the field paralleled the early ones for medicine itself: do not harm anyone (truth telling), do not steal anyone else's patients (another's research), and behave with character (same).[4] When I first started graduate school in 1969, the rules were about being a person of character and not lying: just find the facts, create a reasonable narrative from them, and tell the truth. Thorny ethical concerns were never discussed.

Truth telling, however, especially around the untold histories, was what mattered to us then. I have this vivid memory of a professor saying, for example, "The people of New York State decided to build the Erie Canal," and some waggish grad student asking, "Who decided, professor?" Or I recall querying an eminent historian in my European intellectual history course at New York University who believed the ideas of some obscure French Revolutionary thinker. "That, my dear," he told me in front of 200 others in the lecture course, "is the wrong question."[5] This scholar aside, ethics at that time seemed to be focused on producing "bottom-up history" that asked the "right questions" as determined by the unproblematized social history framework. I also had no idea that any form of health, medical, nursing, or public health history existed. Unhappy with the program I was in, I dropped out, worked as a lefty health journalist and agitator, and thought I was going to be a labor historian, maybe, or perhaps go to public health school.

In January 1975 (after coediting a women's labor history documentary book), I returned to get a PhD, this time at Boston University, in American Studies.[6] As I began my new graduate program, I was taught, in addition to truth telling, that there were two Ps that defined the ethical dilemmas in historical scholarship: plagiarism and presentism. Do not steal other people's ideas (musical pundit Tom Lehrer's lyrics aside), and give history its own past. Do not be like those old Communist Party hack historians I was warned against who wrote to serve a political agenda.[7] The biases of the Cold War liberal historians, of course, were never mentioned, and the concerns with positionality that Barron Lerner raises in his essay were not discussed.

I had to learn about the ethical conundrums of being a historian on my own. The first confrontation I remember came when I did an oral history of a well-known health union labor leader in 1976. To prepare for the work, I read about her differences over a strike with the man who was still her mentor and boss. I sent her the oral history as I had transcribed it, and she changed it to downplay their dispute. Concerned that it was after all her story, I wondered what I also owed the history goddess Clio in my truth telling. I left in what she said, discussed the limits of oral history in my introduction, and cited the article that contradicted her in a footnote.[8]

I also learned that some of the rules that were supposed to cover ethical historical behavior were irrelevant. When I applied for federal money to support my dissertation on the history of nursing, the required institutional review board members wanted me to write a questionnaire for my oral histories. I wrote one up that I knew was ridiculous, since at the time the difference between social science interviews and oral history was not yet clear. I passed inspection. I barely followed the questionnaire, however, since I was responding to what the interviewees were saying, and no one ever checked. Once again, I received no guidance on whatever ethical issues there were here. Ethically, I thought I was on my own. As Marco Ramos's essay makes clear, this is not always a comforting thought.

As oral history developed as a field and I taught it, others started to grapple with these kinds of concerns, as well as what it meant when gender, ethnicity, race, or class differences were present between the historian and their subject.[9] More than that, we learned to consider what it meant to do interviews when silence and secrets were so much a part of women's lives, and especially their reproductive and sexual lives. Did we have the right to contradict our interviewees?[10] The issues I had struggled with in my first interviews increasingly came under scrutiny as many others began to write on these issues and made me feel less isolated.[11] And the work of oral historian Alessandro Portelli helped me to imagine that even if our informants did not tell the truth, the job of the historian was to figure out why the stories were told the way they were.[12]

Somewhat similar questions arose when I worked to turn my 1982 dissertation, with the ungainly, if accurate, title "The Nursing Disorder," into a book: Whose story was I telling? Could I name the class divisions in nursing without insulting nurses? What about racial divisions? At the time I was writing the thesis, I was told that a white woman writing on Black women would never be hired. So I did not cover this obviously crucial historical analysis. Job concerns, not historical accuracy, shaped my intellectual endeavors.[13] I also was repeatedly asked whether I had been a nurse, although no one seemed to query my labor history brothers if they were shoemakers or ironworkers when they wrote on them. I found myself laughingly parodying the old House Un-American Activities Committee pledge, "I am not now nor have I ever been a nurse." I did realize of course that my lack of nursing education limited what I could know or understand. I also believed that asking others with that training to read what I wrote would make it both more ethical and accurate.

As I was finishing graduate school, my friend David Rosner and I became well aware how hard it was to get social history of medicine articles published in the leading journals in our field.[14] As we have written elsewhere, we made a decision to try to edit a book on this topic to stake out an intellectual territory we thought of as "beyond the great doctors."[15] While we saw this primarily as an historiographic and methodological intervention, as "New Lefties" we both came to see what we were doing as a kind of political intervention to consider all the silences and hidden histories of health care that needed to be articulated. We did not at the time consider it an "ethical imperative," although now I think, given the essays in this volume, you might want to articulate it that way as well.[16]

Once my nursing history book was published and my tenure at Wellesley procured, I struggled with and abandoned a number of next book projects. The ethics of positionality was always a key factor, as I had begun to have an interest in nurse Eunice Rivers Laurie and her role in the infamous syphilis study at Tuskegee. I was aware there was an ethical issue: I was a white woman, and this was a loaded and racist study.[17] I asked Darlene Clark Hine, the author of *Black Women in White*, why she had not written much about Rivers in her book.[18] "I'm not concerned about her, you are," she told me. "But I am a white woman," I protested. "How can I write about this?" "I will protect you," she promised, because she believed we should be judged by what we write, not our identities. Not everyone I encountered thought I had the right to do this, however.

Nevertheless, I started. I read the major books and articles, primary and secondary, on the study. I went to see what was in various archives. It became clear that in contrast to the famous definition articulated by my friend Harry Marks that historians "read dead people's mail," I had to speak with people who were alive. I started to have qualms about doing some interviews in Tuskegee because I was writing to set them up from Cambridge (where I was then on sabbatical at the W. E. B. Du Bois Research Institute at Harvard) on Du Bois stationery to Booker T. Washington land, and I did not want anyone to think I was Black. (This was pre-internet, when it was harder to find out.) I thought it might make a difference in people agreeing to an interview, or I did not want to surprise them. If asked, I was open and honest. One potential interviewee asked how she would find me in the Piggly Wiggly Grocery Store parking lot where we agreed to meet. "You cannot miss me," I assured her. "Avis rented me a white car and I'm a white woman." Macon County, which surrounds Tuskegee, is 80 percent African American. She found me. Luckily too, I made friends with the assistant archivist at Tuskegee, whom everyone knew and called to ask if they should speak to me. She vouched for me. I felt as if I had become an anthropologist, not a historian. I had no ethical guidance about how to do any of this work—but being honest about my position seemed the most obvious way forward.

My first effort with this project was to take all the materials I had been gathering to create an edited book on various ways of seeing the study, from a Southern apologia to the most conspiratorial views.[19] My thought was that an ethical way to understand and teach the study was to provide multiple viewpoints. As I began to consider writing a follow-up monograph that looked at how the study happened, who worked on it, and how it was remembered, I had to reach out to do more interviews.[20] It was still hard. I remember I desperately wanted to speak to several of the study's survivors who were then still alive. But I could not figure out how to call them up: "Hi, I am a historian and I want to talk to you about how you got a sexually transmitted disease and the government screwed you over and treated you in racist ways," was accurate but hardly something I could get out of my mouth. Luckily, when I became involved in getting the federal apology for the study at the White House in 1997, I met several of the men and their families there. Calling them up after that was easier and my explanation of what I was working on more thoughtful.

I was still quite worried I would retraumatize the men or their families. This came to a head when I was invited by a family member to give a talk at the Shiloh Missionary Baptist Church in Notasulga, outside Tuskegee, to the congregation, many of whom were relatives of the men in the study. Who was I, a Northern white women, daring to speak to these people about their own family experiences? This ethical problem had never been discussed in my training. My years as a feminist lefty health journalist taught me how to confront arrogant white doctors, but neither that nor my graduate training showed me how to reach those harmed by an immoral research experiment.

Two things saved me, other than Black Southern hospitality, by virtue of which I suspected that even if they thought I was a racist idiot they would not tell me. First, I had just found the photo file of the men from the study, taken by the Public Health Service in the mid-1950s, in the Centers for Disease Control (CDC) records in the Southeast Regional National Archives in Morrow, Georgia. I brought the file with me to Shiloh, and we figured out a way to project the photos on a sheet on the wall at the church. No one had seen them before. I also was saved by my Hebrew school training and my realization of the need for a biblical parable. Jews are required at Passover to tell the story to their children of the forty-year wandering of our ancestors in the desert so it is not forgotten. The study in Tuskegee had lasted for forty years. So I promised the family members at the church as I spoke from their pulpit that I would carry on this lesson and make sure their ancestors and their forty-year sojourn in a medical desert would not be forgotten either. But as other historians have asked, how much do historians owe the dead, and how much and what to the living? Do the dead also have the right to silence? I was not sure.[21]

I faced another dilemma as the men's medical records became available in the same archives, which I then coded to make some quantitative claims and took notes on. If I quoted from them, should I use the men's real names? I made a decision that if they had been public in media accounts about their participation, I would use their name. If not, I made up a name but gave the real record number from the archives. I discussed this with family members in Tuskegee, and they agreed with this approach. No institutional review board oversaw this work. Given Adria L. Imada's essay here, I now can see I was practicing what she calls an "ethics of restraint" without knowing or naming it as such.

As I gave talks about my work, I was constantly struggling with what people thought the study was about, what I could and could not explain, and how I had undertaken my project.[22] There was so much current investment in the meaning of the study that I had to make sure I understood where an audience was and that I was able to speak *with* them rather than *at* them. I was also very aware, as I spoke in Tuskegee and elsewhere, that the descendants of the men and the communities (writ large) they were part of are also bystanders to the study and its trauma. Indeed I could argue that all African Americans are themselves bystanders, in a sense, to the structural violence that underlay the study's very conception.[23] I began to think about what historians have to do to be cognizant of this, especially after the COVID-19 pandemic hit and references to the study in Tuskegee became saliant once again.[24] As Ayah Nuriddin points out in her essay in this volume, the violence that underlies the racism in the history of medicine cannot and should not be denied, but how we tell this story requires all our historical skills and commitment to nuance and contingency.

The ethical concerns I faced with my next, not expected, project were different because they involved thinking more globally and about the United States' control over other countries' health affairs. As I have told the story elsewhere, while I was doing the research for the book about the study in Tuskegee, I found records on a study funded by the National Institutes of Health and given to the Public Health Service carried out between 1946 and 1948 in Guatemala. No one had heard about this study that involved *infecting* sex workers, soldiers, prisoners, and mental hospital patients with a variety of sexually transmitted diseases, including syphilis. It took a while to write this up, and I shared the results with David Sencer (whom I had been interviewing), the former CDC director and a syphilologist, before I gave them as a paper and had them published. Because of Sencer's concerns, my paper, notes, and xeroxes went to the CDC leadership in the spring of 2010.[25] Over the next few months, the story and my work (amazingly to me) went up the chain of command to the White House. Then there was quickly (in government time) a formal apology from the Obama administration to the government of Guatemala on October 1, 2010, for this horrific study. And when the story immediately hit the media, I spent weeks talking to reporters from around the world.

For the next two years, I gave talks on the Guatemala study (and the one in Tuskegee too), and I worked a little with the Presidential Commission for the Study of Bioethical Issues as they did their report on what had happened in Guatemala with much more detail. I was asked to become a historical consultant when a group of attorneys representing Guatemalan family members tried to sue two medical research institutions (Johns Hopkins and the Rockefeller Foundation) and the drug company Bristol Myers Squibb, which supplied the penicillin, for the role of their employees (John Earle Moore, Thomas Parran, Frederick Soper, and others) in making the study happen.[26] It took months of internal struggle before I agreed to do it, because I thought in the end the Guatemalan families deserved their day in court and reparations. I also believed I had enough evidence to show how much of a role these senior physicians had played in making the research happen. I sat through a grueling day of depositions with the institutions' attorneys, who had read everything I had written on the subject and questioned me relentlessly.

The judge, however, did not agree with my position or those of the lawyers on the side of the Guatemalans. In his opinion, for example, I had failed a legal and historical standard. He argued,

> Plaintiffs are therefore left with only the opinion of their historical expert, Dr. Reverby, the historian who uncovered the Guatemala Experiments. Dr. Reverby expressed the opinion that "Moore and others on the Syphilis Section knew what they were doing edged over an ethical boundary" and that Dr. Cutler "had the support of the NIH Syphilis section" and others, including Dr. Moore, "to do what they thought was crucial." J.R. 4783. In support of this opinion, however, Dr. Reverby did not cite any historical documents or records she reviewed but instead relied on the more general opinion that Dr. Moore was "too thorough a scientist to ever have approved this without knowing exactly what was going on" and "would have to have seen the details" to approve it. J.R. 1116. When asked if there is any evidence of what these "details" were, Dr. Reverby agreed that there is no record of them.[27]

In his conclusion, the judge decided that rulings in a "court of law, as opposed to the reasoned judgments of historians, must be grounded in admissible evidence."[28] His position reflects what Stephen Casper writes in his essay about the difference between historical and legal evidence. In this case, it did not end with this judge's ruling: the lawsuit was argued again in December 2023 and was on appeal in the Fourth Circuit, which decided there was not enough link between the physicians in this work to the institutions being sued. I felt that I had failed in this case, not in an ethical sense but historically, because I had not been able to provide enough evidence to be persuasive, and my judgment was not enough. Yet I know we often reconsider historical evidence and come to different conclusions.[29]

While I was unable to persuade a judge that restorative justice through compensation in the courts would work, I do hope my scholarship, talks, and testimony have provided a form of "restorative history" for those harmed at the time, as well as their descendants and community members.[30] As Kylie Smith argues in her essay in this volume, our historical work can serve as a way to "create accountability" in a practice of "reparatory history" even when we cannot control the outcome. As I promised my audience at the Shiloh Missionary Baptist Church, part of our job is to make sure accurate memories of historical experiences are not forgotten.

Finally, as the younger scholars in this volume have demonstrated, the ethical issues for historians in medical, nursing, and health history are nuanced and complicated. My experiences represent a particular moment in time that needs to be historicized and not forgotten. But they do not have to represent the present or the future.

So now that I have weirdly (to me) become an elder in this field, here are my parting thoughts. Plagiarism, presentism, and lying are of course *not* supposed to happen in historical writing. That is the easy part. Our own positionality in relationship to both whom we write about and whom we write for needs to be discussed too. This should not stop us from taking up subjects for which we have passion and training, but it should make us mindful of what it might mean to do this kind of work, and we should struggle to find the best way to name that struggle. I think we have to balance the demands of historical accuracy with the needs of the living as best we can. We also have to be concerned with our self-care as scholars, as much as we care for those whose histories we try to write.

Finally, in the context of a political climate where the kind of history we are all writing has come under attack, those of us in somewhat protected positions have a duty to stand up, protect our younger colleagues, and demonstrate what ethical and accurate history looks like. People's lives are on the line here, and we ought to be active as historians in making this clear, for example, in this current political moment with the resurrection of nineteenth-century abortion laws and the attack on science and health care.[31] In that sense too, we should be "ordered to care" and to think as much ethically as we are taught to write historically.

NOTES

I thank editors Kylie Smith and Courtney Thompson, and the anonymous reviewers, for their extremely helpful editorial comments. I have faith our field is in good hands, given the thoughtful essays in this volume and the model for the process of creating that this book has developed.

1. Tom Lehrer, "Lobachevsky," 1951, https://www.lyrics.com/sublyric /5746/Tom+Lehrer/Lobachevsky.

2. James H. Sweet, "Is History History? Identity Politics and Teleologies of the Present," *Perspectives on History*, August 17, 2022, https://www .historians.org/research-and-publications/perspectives-on-history/september -2022/is-history-history-identity-politics-and-teleologies-of-the-present.

3. Antoon De Baets, *Responsible History* (New York: Berghahn Books, 2008), 191.

4. Robert Baker, *Before Bioethics: A History of American Medical Ethics from the Colonial Period to the Bioethics Revolution* (New York: Oxford University Press, 2013).

5. I used to tell this story to my students to explain the difference between intellectual and social history. And the fact that Wellesley gave me a chair in the History of Ideas must have made my former professor, the eminent intellectual historian Frank E. Manuel, roll over in his grave.

6. Roslyn Baxandall, Linda Gordon, and Susan Reverby, eds., *America's Working Women: A Documentary History* (New York: Random House, 1976). In a "maybe this could happen only in New York City in the 1970s" story, I sold the idea for the book to Toni Morrison, who was then a Random House editor. I had just finished my MA, Ros was an ABD, and Linda was then a Russian historian.

7. Dutifully, I only worked with Howard Zinn on picket lines during clerical and faculty strikes but never had him as a teacher.

8. Susan Reverby, "From Aide to Organizer: The Oral History of Lillian Roberts," in *Women in America: A History*, ed. Carol Berkin and Mary Beth Norton (Boston: Houghton Mifflin, 1978), 289–317.

9. See, for example, the influential essay by Catherine Kohler Riessman, "When Gender Is Not Enough: Women Interviewing Women," *Gender and Society* 1, no. 2 (June 1987): 172–207.

10. Katherine Borland, "'That's Not What I Said': Interpretive Conflict in Oral Narrative Research," in *Women's Words: The Feminist Practice of*

Oral History, ed. Sherna Berger Gluck and Daphne Patai (New York: Routledge, 1991), 63–76.

11. Susan H. Armitage, Patricia Hart, and Karen Weatherman, eds., *Women's Oral History: The Frontiers Reader* (Omaha: University of Nebraska Press, 2002).

12. Alessandro Portelli, *The Death of Luigi Trastulli and Other Stories* (Albany: State University of New York Press, 1991).

13. Susan M. Reverby, *Ordered to Care: The Dilemma of American Nursing* (New York: Cambridge University Press, 1987).

14. David Rosner, "Tempest in a Test Tube: Medical History and the Historian," *Radical History Review* 26 (1982): 166–171.

15. Susan M. Reverby and David Rosner, "Beyond 'the Great Doctors,'" in *Health Care in America: Essays in Social History,* ed. Susan M. Reverby and David Rosner (Philadelphia: Temple University Press, 1979), 3–16; Susan M. Reverby and David Rosner, "'Beyond the Great Doctors' Revisited: A Generation of the 'New' Social History of Medicine," in *Locating Medical History: The Stories and Their Meanings*, ed. Frank Huisman and John Harley Warner (Baltimore: Johns Hopkins University Press, 2004), 167–193.

16. I am grateful to Kylie Smith and Courtney Thompson, and the numerous other essayists in this volume, for thinking of our intellectual intervention as an "ethical imperative" as well.

17. I had gone to graduate school the first time in 1969 to learn Black history and the politics of racism and white supremacy. When I came back to do my PhD six years later, I was told that white people writing about Black history were not getting hired. The field of "whiteness studies" and critiques of white supremacy were just beginning anew, and they started to expand after I got my PhD in 1982. My dissertation became the book Reverby, *Ordered to Care*.

18. Darlene Clark Hine, *Black Women in White* (Indianapolis: Indiana University Press, 1989).

19. Susan M. Reverby, ed., *Tuskegee's Truths: Rethinking the Tuskegee Syphilis Study* (Chapel Hill: University of North Carolina Press, 2000).

20. This work became Susan M. Reverby, *Examining Tuskegee: The Infamous Syphilis Study and Its Legacy* (Chapel Hill: University of North Carolina Press, 2009).

21. For a discussion of this issue, see De Baets, *Responsible History*, esp. 111–144, on the "duties of the living to the dead."

22. Susan M. Reverby, "Ethical Failures and History Lessons: The U.S. Public Health Service Research Studies in Tuskegee and Guatemala," *Public Health Reviews* 34 (January 2013): 1–18.

23. Susan M. Reverby, "Compensation and Reparations for Victims and Bystanders of the U.S. Public Health Service Research Studies in Tuskegee and Guatemala: Who Do We Owe What?," *Bioethics* 34, no. 9 (November 2020): 893–898.

24. On the constant referencing to the study in Tuskegee, see Evelynn M. Hammonds and Susan M. Reverby, "Taking a Medical History: COVID,

'Mistrust,' and Racism," *Mudsill* 1, no. 7 (April 15, 2021), https://themudsill .substack.com/p/the-mudsill-vol-1-no-7.

25. Susan M. Reverby, "'Normal Exposure' and Inoculation Syphilis: A PHS 'Tuskegee' Doctor in Guatemala, 1946–48," *Journal of Policy History* 23 (Winter 2011): 6–28.

26. Estate of Alvarez v. The Johns Hopkins University, 598 F. Supp. 3d 301 (D. Md. 2022), https://casetext.com/case/estate-of-alvarez-v-the-john -hopkins-univ.

27. *Id.* at 341–342.

28. *Id.* at 347.

29. Professor Susan E. Lederer was the historical consultant for Hopkins and Rockefeller in this lawsuit, and she came to historical conclusions that differed from mine.

30. Susan M. Reverby, "Restorative Justice and Restorative History for Sexually Transmitted Disease Inoculation Experiments in Guatemala," *American Journal of Public Health* 106, no. 7 (July 2016): 1163–1164.

31. Pam Belluck, "The History behind Arizona's 160-Year-Old Abortion Ban," *New York Times*, April 11, 2024.

Coda

A Report, a Turn

COURTNEY E. THOMPSON

As we were in the process of completing our revisions to our essays in this volume, I stumbled across a strange thing quite by accident, in a footnote on a thirty-year-old article I was reading while prepping to teach a graduate class. The note, in Guenter Risse and John Harley Warner's classic essay on patient records in medical history, referred briefly to the "recently approved code of ethics of the American Association for the History of Medicine."[1] *What code of ethics?* I wondered. I had never heard of such a thing. I went digging.

There was no mention of this code of ethics anywhere on the current AAHM website that I could find. Nor was there any mention of this code on past versions of the AAHM website going back to 1999, when the first snapshots of the site were captured by the Internet Archive. I next explored old issues of the *Bulletin of the History of Medicine*, the journal of the society, starting with the date of publication of Risse and Warner's essay and moving back. Luckily, there was a six-page "Report of the Committee on Ethical Codes" published in winter 1991.[2]

Though it was not quite the "code" promised by the Risse and Warner footnote, the report is nonetheless quite revealing as to the ethical assumptions and priorities of the field, at least in the United States, more than thirty years ago. Half of the document is devoted to the dangers of plagiarism and how to defend against it (or to penalize offenders). Best practices for oral history are addressed in a page and a half, with specific guidelines adapted from those of the Oral History Association and the Society for History in the Federal Government, in consultation with scholars of the Organization of American Historians and the Society of American Archivists, influenced as well by the American

Historical Association's Statement on Standards of Professional Conduct. Another page and a half, as the introduction to the report, stresses key values for ethics, highlighting integrity, access to sources, intellectual diversity (with a nod to the importance of intellectual freedom), and "standards of civility" around discourse. A single paragraph of this report addresses patient records.

What this report does not address is just as revealing. Aside from a broad gesture in the opening paragraph, little attention is paid to the ethical responsibilities of the health historian in the classroom, to our students, to our publics, or to descendant communities. The focus of these principles is on historical research practices and outputs, and the chief responsibilities seem to be to the profession and our colleagues. In an odd way, this echoes the preoccupations of the original Hippocratic oath, which did not—despite the widespread assumption—declare that a physician should first "do no harm" to patients but instead first and foremost pledged fealty to the gods and then to the profession and its practitioners (patients were a distant third concern).[3] "Access to sources," for example, seems to be principally about making these sources available to other scholars.

The main focus, however, is on plagiarism, which is not surprising. As both Susan Reverby and I observed in this book, plagiarism is often the only point of ethics that emerges in graduate classroom discussions (or with undergraduates, for that matter). The harms of plagiarism to the "pursuit of truth" are emphasized in this report, as is the "obligation to oppose deception actively."[4] Other possible "harms" and "obligations" are not addressed.

The sole paragraph on the "special problem" of patient records (written before the advent of Health Insurance Portability and Accountability Act restrictions) argues that "standards of common sense and good judgment should guide the researcher."[5] The report focuses on confidentiality, especially anonymizing or coding patient records, and particularly in cases of "sensitive subjects" like sexually transmitted disease and genetic conditions like sickle cell. The ethical considerations with regard to patient records end there.

While this was a report produced specifically for AAHM, a US-based organization, reading this report in the context of completing this book is suggestive of the many shifts that have occurred in the past thirty years within our field, broadly speaking. Many of the key ideas present in our book—reparatory history, decolonization, accessibility, social justice, and so forth—are not present in the report. Likewise, there is no chapter in this volume on plagiarism, and oral history, while addressed in various chapters, is treated not as a separate ethical problem but as a component of other ethical concerns, such as capturing and representing the voice of the patient ethically in public history, as in Britt Dahlberg and Jessica Martucci's essay.

The report is an odd artifact and unfortunately something of a black box. I reached out to AAHM leaders, past and present, and it has proved difficult to find meeting minutes or memories that would shed light on what led to the formation of this committee and production of this report. Was there some ongoing debate or controversy within the field that had prompted these developments, or were external pressures—perhaps relating to the field of history as a whole or the politics of American health care in the early 1990s—directing this work? The report is silent on this point, and the surviving members of the committee and AAHM leadership of this period have few memories of this time that they were able to share with me.[6]

Equally intriguing is that the report itself seems to have languished in obscurity. Whether or not it was intended as a true "code of ethics," it seems to have never been used as such and all but forgotten. According to (the admittedly imperfect) Google Scholar, this document has never been cited (the discursive footnote I originally found, recall, did not include a citation to the published report). Even Susan Lawrence's *Privacy and the Past*, which addresses ethics and the history of medicine, does not cite or allude to it, even when discussing other codes of ethics.[7] Senior members of AAHM with whom I discussed this document were surprised that such a thing had ever existed. If this was intended to be a guide to member ethics, it was not one that was widely circulated, known, or used.

It is not strange that the priorities of the report writers and AAHM council who approved the report were so different thirty years ago from those of our collective. Nor is it surprising that this "code of ethics" was not widely adopted. But the field has changed substantially: we have different concerns that extend beyond those of plagiarism and access to archival sources.

Do Less Harm does not amount to a new code of ethics for health historians. Instead, we have very intentionally offered case studies of dilemmas and key questions for the reader to contemplate. We do not call for a code per se but rather an integration of ethical considerations into *all* health historical work, from conception to completion. Ethics is not something to consider only in a small set of specific circumstances but instead should be a constituent element of all of our labors.

The 1991 report seems to have been an end to a conversation, rather than the start of one. While such codes can be useful, the seemingly definitive mark of them seems to preclude evolution and further development of ethical principles. We hope that this volume will be a beginning to a much broader conversation that does not end with this publication. We are thus calling not for a new code or a revision to the report of 1991 but for a more profound change: an ethical turn in health history.

NOTES

1. Guenter B. Risse and John Harley Warner, "Reconstructing Clinical Activities: Patient Records in Medical History," *Social History of Medicine* 5, no. 2 (1992): 189n19.

2. Kenneth F. Kiple et al., "American Association for the History of Medicine: Report of the Committee on Ethical Codes," *Bulletin of the History of Medicine* 65, no. 4 (Winter 1991): 565–570.

3. The translation of the Hippocratic oath with which I am most familiar is in David J. Rothman, Steven Marcus, and Stephanie A. Kiceluk, eds., *Medicine and Western Civilization* (New Brunswick, NJ: Rutgers University Press, 1995), 261–262.

4. Kiple et al., "American Association for the History of Medicine: Report of the Committee on Ethical Codes," 568, 569.

5. Kiple et al., 567.

6. I was able to track down one of the four original authors and committee members, Samuel Thielman, who was kind enough to speak with me on the phone. Unfortunately, he has no memory of how the report or the committee itself came together. In fact, he told me that the first time he ever saw the report on which he was listed as an author was when I sent it to him in preparation for our discussion, "which is an ethical concern right there," in his words. Charles Rosenberg, who was then the vice president of AAHM, similarly had no recollections of how this report came to be. The then president, John Burnham, and two of the four members of the committee have since passed; I was unable to find contact information for the fourth committee member. Samuel Thielman, in discussion with the author, April 9, 2024; Charles Rosenberg, email message to the author, March 5, 2024.

7. Susan C. Lawrence, *Privacy and the Past: Research, Law, Archives, Ethics* (New Brunswick, NJ: Rutgers University Press, 2016).

Ethical Questions
for Health Historians to Consider

Each of the authors contributed some key questions that have informed their own approach. We have organized them here by book sections, although some are relevant across areas. We hope you find them provocative and helpful in your own research, writing, practice, and teaching.

Part I: The Historian

If ethical research conduct is a long-term process, not a single event or consent form, what implications does this hold for your research project and for your career as a historian?

Why do some scholars now think it should be mandatory for historians to declare their positionality in academic papers? Do you agree?

How and why should historians of medicine and health engage with debates about racism and reparations?

As we work toward more ethical scholarship, how will we engage with questions of justice in our work?

When we claim we value diversity, equity, and inclusion in the history of medicine, what do we mean? How can we move beyond such rhetoric into action?

How can we better build access for those who are not well represented in the field?

Part II: Archives and Museums

How can you balance using documents written from an institutional perspective with encouraging an empathetic view of the patient when you do not have their perspective?

How can you use disability justice principles to help inform your ethical practices when dealing with marginalized populations?

How might a shift toward more ethical or reflective research practices by historians affect research repositories (archives, museums, and libraries)?

How can different communities beyond those in academic institutions shape the ethics surrounding human remains collection and display?

How can medical heritage collections address the shifting politics and ethical concerns surrounding human remains, considering historical collecting practices and the lack of patient voice and consent?

What are some issues you should consider when collecting or buying medical or health artifacts?

Part III: Research

What role should legal records play in the writing of medical history? What is the best approach to critically evaluating them as sources?

Why does an institutional review board or ethics committee exempt oral history projects from ethical review? Do you agree that they should be?

If the institutional review board or ethics committee exempts our research, how can we create alternative spaces or processes to explore the deeper ethics of our work?

How can historians ethically document histories of scientific racism and medical violence?

How can historians center and amplify the voices of marginalized peoples?

What are some ethical conflicts of interest involved in participating in legal cases? Does the answer change depending on whom you might be representing?

How does thinking through a global, decolonial lens help us think ethically about the histories of medicine and health that we produce?

Part IV: Writing

How and when should historians use the names of deceased people in their accounts?

Is it ever acceptable to use clinical photographs in your writing? What are some ways you can approach this issue with consideration for the people portrayed? How would you feel if that person were you?

When and why should you use person-first language? When and why do you plan to use diagnostic categories in your writing?

Whom are you including in your citations? Whom are you excluding? What do these choices signal about your scholarly concerns and community?

As a peer reviewer or editor, how can you contribute to making a project better rather than acting as a gatekeeper?

What ethical obligations do you have when writing for public spaces, especially online? When writing for the public, whom are you writing to, for, and with? How do your results line up with your goals and intentions?

What power dynamics and whose assumptions are framing your work? Is there space or need to shift the lines between scholar, subject, and audience? How would we go about doing this ethically?

Part V: Teaching

How should you approach using specimens and museum artifacts in your pedagogical practice, particularly if these are unethically sourced or displayed?

What is your approach to teaching difficult topics? What practices can you employ in order to balance comfort and discomfort in the classroom?

How can you use sensitive images in the classroom in a way that not only avoids triggering students but also opens a discussion on the ethical responsibilities of the historian?

How can you center diversity in positionality among the work of historians we use in teaching while still honoring differences of opinion?

How can you support your students in thinking about ethics if they are working with other mentors who view it as frivolous?

What are the conceptual goals of feminist critical pedagogy? How does feminist critical pedagogy assist future health professionals

to learn how to thrive in unfamiliar environments and meaningfully confront difference?

What are some examples of learning spaces that engage with ethical issues of expansion, inclusion, and justice in health care
education?

How can we best make sure the ethical issues raised in this volume get shared with students and faculty alike?

Recommended Readings

In addition to the reading list from our original workshop, every author in this volume also suggested a few key readings that informed their approach to ethics and health history. This list is not a comprehensive bibliography of all possible works that touch on these subjects but instead a starting point for our work (and perhaps yours).

Bacopoulos-Viau, Alexandra, and Aude Fauvel. "The Patient's Turn: Roy Porter and Psychiatry's Tales, Thirty Years On." *Medical History* 60, no. 1 (2016): 1–18.

Bate, Jason. *Photography in the Great War: The Ethics of Emerging Medical Collections from the Great War.* London: Bloomsbury Academic, 2021.

Beauchamp, Tom L. "The Nature of Applied Ethics." In *A Companion to Applied Ethics,* edited by R. G. Grey and Christopher Heath Wellman, 1–16. Hoboken, NJ: John Wiley and Sons, 2005.

Biernoff, Suzannah. "Medical Archives and Digital Culture." *Photographies* 5, no. 2 (2012): 179–202.

Brake, Elizabeth. "Rebuilding after Disaster: Inequality and the Political Importance of Place." *Social Theory and Practice* 45, no. 2 (2019): 179–204.

Burgess, Hana, Donna Cormack, and Papaarangi Reid. "Calling Forth Our Pasts, Citing Our Futures: An Envisioning of Kaupapa Māori Citational Practice." *MAI Journal* 10, no. 1 (2021): 57–67.

Butcher, Kacie Lucchini. "More Questions Than Answers: Interrogating Restricted Access in the Archives." *Journal of the History of the Behavioral Sciences* 58, no. 2 (2022): 223–231.

Campbell, Nancy D., and Laura Stark. "Making Up 'Vulnerable' People: Human Subjects and the Subjective Experience of Medical Experiment." *Social History of Medicine* 28, no. 4 (2015): 825–848.

Canaday, Margot. *Queer Career: Sexuality and Work in Modern America.* Princeton, NJ: Princeton University Press, 2023.

Canadian Institutes of Health Research, Natural Sciences and Engineering Research Council of Canada, and Social Sciences and Humanities

Research Council of Canada. *Tri-council Policy Statement: Ethical Conduct for Research Involving Humans.* Government of Canada, December 2022. https://ethics.gc.ca/eng/policy-politique_tcps2-eptc2_2022.html.

Carmen Salazar, María del. "A Humanizing Pedagogy: Reinventing the Principles and Practice of Education as a Journey toward Liberation." *Review of Research in Education* 37, no. 1 (2013): 121–148.

Carniel, Jessica. "[Insert Image Here]: A Reflection on the Ethics of Imagery in a Critical Pedagogy for the Humanities." *Pedagogy, Culture and Society* 26, no. 1 (2018): 141–155.

Chakrabarty, Dipesh. *Habitations of Modernity: Essays in the Wake of Subaltern Studies.* Chicago: University of Chicago Press, 2002.

Chibber, Vivek. *Postcolonial Theory and the Specter of Capital.* London: Verso, 2013.

Crane, Susan. "Choosing Not to Look." *History and Theory* 47, no. 3 (2008): 309–330.

Czech, Herwig, Christiane Druml, and Paul Weindling, eds. "Medical Ethics in the 70 Years after the Nuremberg Code, 1947 to the Present." *Wiener klinische Wochenschrift* 130, Suppl. 3 (2018): 159–253.

DasGupta, Sayantani. "Teaching Medical Listening Through Oral History." *Literature Arts and Medicine,* July 24, 2008. https://medhum.med.nyu.edu/magazine/archives/126.

De Baets, Antoon. *Responsible History.* New York: Berghahn Books, 2008.

Drake, Jarrett M. "Diversity's Discontents: In Search of an Archive of the Oppressed." *Archives and Manuscripts* 47, no. 2 (2019): 270–279.

Elliott, Carl. *Better Than Well: American Medicine Meets the American Dream.* New York: Norton, 2004.

Ellsworth, Elizabeth. "Why Doesn't This Feel Empowering? Working through the Repressive Myths of Critical Pedagogy." *Harvard Educational Review* 59, no. 3 (September 1, 1989): 297–325.

Farmer, Ashley. "Archiving While Black," *Black Perspectives* (blog), June 18, 2018. https://www.aaihs.org/archiving-while-black/.

Figlio, Karl. *Remembering as Reparation: Psychoanalysis and Historical Memory.* London: Palgrave Macmillan, 2017.

Fischer, Michael M. J. *Emergent Forms of Life and the Anthropological Voice.* Durham, NC: Duke University Press, 2003.

Fisher, Carl Erik. *The Urge: Our History of Addiction.* New York: Penguin, 2022.

Fixico, Donald L. "American Indian History and Writing from Home: Constructing an Indian Perspective." *American Indian Quarterly* 33, no. 4 (Fall 2009): 553–560.

Froeyman, Anton. *History, Ethics, and the Recognition of the Other: A Levinasian View on the Writing of History.* New York: Routledge, 2015.

Fuentes, Marisa J. *Dispossessed Lives: Enslaved Women, Violence, and the Archive.* Philadelphia: University of Pennsylvania Press, 2016.

Gabel, Susan. "Some Conceptual Problems with Critical Pedagogy." *Curriculum Inquiry* 32, no. 2 (2002): 177–201.

Gannon, Kevin. *Radical Hope: A Teaching Manifesto.* Morgantown: West Virginia University Press, 2020.

Garland-Thomson, Rosemarie. *Staring: How We Look.* New York: Oxford University Press, 2009.

Gazi, Andromache. "Exhibition Ethics—an Overview of Major Issues." *Journal of Conservation and Museum Studies* 12, no. 1 (2014): article 4. https://doi.org /10.5334/jcms.1021213.

Gebhard, Amanda, Sheelah McLean, and Verna St. Denis, eds. *White Benevolence: Racism and Colonial Violence in the Helping Professions.* Black Point, Nova Scotia: Fernwood, 2022.

Giroux, Henry A. *The Violence of Organized Forgetting: Thinking beyond America's Disimagination Machine.* San Francisco: City Lights, 2014.

Goldberg, Daniel S. "The Shadows of Sunlight: Why Disclosure Should Not Be a Priority in Addressing Conflicts of Interest." *Public Health Ethics* 12, no. 2 (2019): 202–212.

Gur-Ze'ev, Ilan. "Feminist Critical Pedagogy and Critical Theory Today." *Journal of Thought* 40, no. 2 (2005): 55–72.

Hall, Catherine. "Doing Reparatory History: Bringing 'Race' and Slavery Home." *Race and Class* 60, no. 1 (2018): 3–21.

Hartman, Saidiya V. *Scenes of Subjection: Terror, Slavery, and Self-Making in Nineteenth-Century America.* New York: Oxford University Press, 1997.

Hartman, Saidiya V. "Venus in Two Acts." *Small Axe* 12, no. 2 (2008): 1–14.

Holmes, Andrew G. D. "Researcher Positionality—a Consideration of Its Influence and Place in Qualitative Research—a New Researcher Guide." *Shanlax International Journal of Education* 8, no. 4 (2020): 1–10.

Hong, Cathy Park. *Minor Feelings: An Asian American Reckoning.* New York: One World, 2020.

hooks, bell. *Teaching to Transgress.* New York: Routledge, 1994.

Huisman, Frank, and John Harley Warner, eds. *Locating Medical History: The Stories and Their Meanings.* Baltimore: Johns Hopkins University Press, 2004.

Jordanova, Ludmilla. *History in Practice.* London: Bloomsbury Academic, 2019.

Jordanova, Ludmilla. "The Social Construction of Medical Knowledge." *Social History of Medicine* 8, no. 3 (1995): 361–382.

Keene, Rochelle, and Julie Parle. "Museums, Archives and Medical Material in South Africa: Some Ethical Considerations." *South African Museums Association Bulletin* 37, no. 1 (2015): 6–16.

Kilroy-Marac, Katie, and Kassandra Spooner-Lockyer. "Ten Things about Ghosts and Haunting." *Anthropology News*, October 18, 2021. https://www .anthropology-news.org/articles/ten-things-about-ghosts-and-haunting/.

Kincheloe, Joe L. "Critical Pedagogy in the Twenty-First Century: Evolution for Survival." *Counterpoints* 422 (2012): 147–183.

Klein, Kerwin Lee. "On the Emergence of Memory in Historical Discourse." *Representations* 69 (2000): 127–150.

Klugman, Craig M., and Erin Gentry Lamb. *Research Methods in Health Humanities.* Oxford: Oxford University Press, 2019.

LaCapra, Dominick. *Writing History, Writing Trauma*. Baltimore: Johns Hopkins University Press, 2014.

Laite, Julia. "The Emmet's Inch: Small History in a Digital Age." *Journal of Social History* 53, no. 4 (2020): 963–989.

Lawrence, Susan C. *Privacy and the Past: Research, Law, Archives, Ethics*. New Brunswick, NJ: Rutgers University Press, 2016.

Mancuso, Rebecca. "The Finger Saga: One Museum's Quest to Turn the Macabre into the Meaningful." *Public Historian* 40, no. 2 (2018): 23–42.

Mbembe, Achille. "The Power of the Archive and Its Limits." In *Refiguring the Archive*, edited by Carolyn Hamilton, Verne Harris, Jane Taylor, Michele Pickover, Graeme Reid, and Razia Saleh, 19–26. Dordrecht: Kluwer Academic, 2002.

McCallum, Mary Jane Logan. "Starvation, Experimentation, Segregation, and Trauma: Words for Reading Indigenous Health History." *Canadian Historical Review* 98, no. 1 (2017): 96–113.

Mills, Mara, and Rebecca Sanchez, eds. *Crip Authorship: Disability as Method*. New York: New York University Press, 2023.

Mirzoeff, Nicholas. *White Sight: Visual Politics and Practices of Whiteness*. Cambridge, MA: MIT Press, 2023.

Moeller, Frank. "The Looking/Not Looking Dilemma." *Review of International Studies* 35, no. 4 (2009): 781–794.

Moraga, Cherríe, and Gloria Anzaldúa, eds. *This Bridge Called My Back: Writings by Radical Women of Color*. 4th ed. Albany: State University of New York Press, 2015.

Mott, Carrie, and Daniel Cockayne. "Citation Matters: Mobilizing the Politics of Citation toward a Practice of 'Conscientious Engagement.'" *Gender, Place and Culture* 24, no. 7 (2017): 954–973.

Nair, Aparna, and Kylie M. Smith. "We're Historians of Disability. What We Just Found on eBay Horrified Us." *Slate*, July 21, 2022. https://slate.com/technology/2022/07/vintage-asylum-records-found-on-ebay-history-of-disability.html.

Neumeyer, Joy. "Darkness at Noon: On History, Narrative, and Domestic Violence." *American Historical Review* 126, no. 2 (2021): 700–707.

Nicholas, Jane. "A Debt to the Dead? Ethics, Photography, History, and the Study of Freakery." *Social History* 47, no. 93 (2014): 139–155.

Peña, Lorgia García. *Community as Rebellion: A Syllabus for Surviving Academia as a Woman of Color*. Chicago: Haymarket Books, 2022.

Porter, Roy. "The Patient's View: Doing Medical History from Below." *Theory and Society* 14, no. 2 (March 1985): 175–198.

Price, Margaret. *Crip Spacetime: Access, Failure, and Accountability in Academic Life*. Durham, NC: Duke University Press, 2024.

Reisman, Anna. "Should Doctors Write about Patients?" *Atlantic*, February 18, 2015. https://www.theatlantic.com/health/archive/2015/02/should-doctors-write-about-their-patients/385296.

Reverby, Susan M. "Ethical Failures and History Lessons: The U.S. Public Health Service Research Studies in Tuskegee and Guatemala." *Public Health Reviews* 34, no. 1 (2012): 1–18.

Reverby, Susan M., and David Rosner. "Beyond 'the Great Doctors.'" In *Health Care in America: Essays in Social History*, edited by Susan M. Reverby and David Rosner, 3–16. Philadelphia: Temple University Press, 1979.

Reverby, Susan M., and David Rosner. "'Beyond 'the Great Doctors' Revisited: A Generation of the 'New' Social History of Medicine." In *Locating Medical History*, edited by Frank Huisman and John Harley Warner, 167–193. Baltimore: Johns Hopkins University Press, 2004.

Roelcke, Volker, and Giovanni Maio. *Twentieth Century Ethics of Human Subjects Research: Historical Perspectives on Values, Practices, and Regulations*. Stuttgart: Franz Steiner, 2004.

Sadowsky, Jonathan, and Kylie Smith. "Reflections on the Use of Patient Records: Privacy, Ethics and Reparations in the History of Psychiatry." *Journal of the History of the Behavioral Sciences* 60, no. 1 (2023): e22260.

Satia, Priya. "The Presentist Trap." *Perspectives on History*, September 7, 2022. https://www.historians.org/research-and-publications/perspectives-on -history/october-2022/responses-to-is-history-history.

Scott, David. "Preface: A Reparatory History of the Present." *Small Axe* 21, no. 1 (2017): vii–x.

Scott, Joan Wallach. "The Evidence of Experience." *Critical Inquiry* 17, no. 4 (Summer 1991): 773–797.

Scott, Joan Wallach. *On the Judgment of History*. New York: Columbia University Press, 2020.

Shalvey, Aisling. "Naming the Invisible Patient." *Polyphony*, June 18, 2021. https://thepolyphony.org/2021/06/18/naming-the-invisible-patient-a -historical-dilemma/.

Sharpe, Jenny. *Immaterial Archives: An African Diaspora Poetics of Loss*. Evanston, IL: Northwestern University Press, 2020.

Sheppard, Maia G. "Creating a Caring Classroom in Which to Teach Difficult Histories." *History Teacher* 43, no. 3 (2010): 411–426.

Simmons, LaKisha Michelle. "Black Women Authors in the Journal of American History." *Journal of American History*, 2021, 1–8. https://academic.oup .com/jah/pages/black-women-authors-in-the-jah.

Smith, Linda Tuhiwai. *Decolonizing Methodologies: Research and Indigenous Peoples*. London: Bloomsbury, 2021.

Stoler, Ann Laura. *Along the Archival Grain: Epistemic Anxieties and Colonial Common Sense*. Princeton, NJ: Princeton University Press, 2008.

Te Hennepe, Mieneke. "Private Portraits of Suffering on Stage: Curating Clinical Photographic Collections in the Museum Context." *Science Museums and Research* 5 (2016). https://dx.doi.org/10.15180/160503.

Titchkosky, Tanya. *The Question of Access: Disability, Space, Meaning*. Toronto: University of Toronto Press, 2011.

Topouzova, Lilia. "On Silence and History." *American Historical Review* 126, no. 2 (2021): 685–699.

Trouillot, Michel-Rolph. *Silencing the Past: Power and the Production of History.* Boston: Beacon, 1995.

Wakely, Helen, and Carly Dakin. "Assessing the Sensitivity of Images in Research Collections: A New Approach at the Wellcome Library." *Journal of Visual Communication in Medicine* 38 (2015): 51–60.

Washington, Harriet. *Medical Apartheid: The Dark History of Medical Experimentation on Black Americans from Colonial Times to the Present.* New York: Doubleday, 2006.

Weiler, Kathleen. "Freire and a Feminist Pedagogy of Difference." In *Debates and Issues in Feminist Research and Pedagogy: A Reader,* edited by Janet Holland, Maud Blair, and Sue Sheldon, 23–44. Clevedon, UK: Multilingual Matters in association with the Open University, 1995.

Weld, Kirsten. *Paper Cadavers: The Archives of Dictatorship in Guatemala.* Durham, NC: Duke University Press, 2014.

Wysong, Lori. "What Would It Mean to Decolonize the Curriculum?" *Hindsights* (blog), October 2, 2020. https://medium.com/hindsights/what-would-it -mean-to-decolonize-the-curriculum-4fcedbe781d1.

Yang, Tongjin. "Is There an Identity Crisis in Environmental Ethics?" *Frontiers of Philosophy in China* 12, no. 2 (2017): 195–206.

Zembylas, Michalinos. "'Pedagogy of Discomfort' and Its Ethical Implications: The Tensions of Ethical Violence in Social Justice Education." *Ethics and Education* 10, no. 2 (2015): 163–174.

STEPHEN T. CASPER is Professor of History at Clarkson University. He is currently completing a monograph titled *Punch Drunk and Dementia: A Cultural History of Concussion, 1850–Present.* He is serving as a plaintiff's expert in concussion litigation pending against the National Collegiate Athletics Association, the Rugby League, the Rugby Union, and the National Hockey League.

CLAIRE D. CLARK is Associate Professor of Behavioral Science and History at the University of Kentucky. She is the author of *The Recovery Revolution: The Battle over Addiction Treatment in the United States* (Columbia University Press, 2017).

MICHAELA CLARK is a Research Associate at the Department of the History, Philosophy and Ethics of Medicine at Heinrich Heine University as well as an Honorary Research Fellow at the Centre for the History of Science, Technology and Medicine at the University of Manchester. Her doctoral research draws on training in visual culture studies and focuses on the history and representational politics of clinical photography in the context of twentieth-century Cape Town, South Africa.

BRITT DAHLBERG is an anthropologist and public humanities scholar. Her ethnographic work has explored life amid environmental risk and uncertainty, as well as how scientific and medical practices unfold in the specific contexts of people's lives. She holds a PhD in anthropology from the University of Pennsylvania and currently works as the Director of Research in the Center for Humanism at Cooper Medical School of Rowan University.

JESS DILLARD-WRIGHT (she/they) is a nurse and midwife working at the University of Massachusetts Amherst Elaine Marieb College of

Nursing where she/they are the Associate Dean for Equity and Inclusion. Her/their scholarship resides in the confluence of health care, activism, history, and philosophy. She/they are coeditor of *Nursing a Radical Imagination: Moving from Theory and History to Action and Alternate Futures* (Routledge, 2022).

MELISSA GRAFE is the Head of the Medical Historical Library at Yale School of Medicine and joined Yale University in 2011 as the John R. Bumstead Librarian for Medical History at the Harvey Cushing/John Hay Whitney Medical Library. She received her PhD in the history of medicine from Johns Hopkins University in 2009 and was a Council of Library and Information Resources postdoctoral fellow at Lehigh University Library.

ADRIA L. IMADA was born and raised in Honolulu, Hawai'i. She is Professor of History at the University of California, Irvine, where she also teaches medical humanities. She is the author of two award-winning books: *Aloha America: Hula Circuits through the U.S. Empire* (Duke University Press, 2012) and *An Archive of Skin, an Archive of Kin: Disability and Life-Making during Medical Incarceration* (University of California Press, 2022).

KATRINA JIRIK lives in St. Paul, Minnesota. She is an independent scholar focusing on issues surrounding disability. She received her degree from the University of Minnesota's Program in the History of Science, Technology, and Medicine in 2019. She has given lectures regarding the intersection of disability and health care to diverse medical communities. Her book, *Living in the Abyss*, a look at the internal experience of living with Complex PTSD, was published in June 2023 by Kelsay Press.

ANTOINE S. JOHNSON is Assistant Professor at the University of California, Davis, where he is working on his book manuscript examining Black AIDS activism in the Bay Area. He earned his PhD in 2022 from the University of California, San Francisco. His research interests include HIV/AIDS activism, anti-Black racism in medicine, hip-hop culture, and twentieth-century US history.

CORNELIA LAMBERT is a historian of science and medicine. Her research interests include Scottish moral philosophy, the histories of looking and seeing in social contexts, and the history of public display of children and families. She teaches at Oglethorpe University in Atlanta.

BARRON H. LERNER is Professor of Medicine and Population Health at the New York University Grossman School of Medicine. He is the author of five books on the history of medicine and public health, including *The Breast Cancer Wars: Hope, Fear and the Pursuit of a Cure in Twentieth-Century America,* which won the William H. Welch Medal from the American Association for the History of Medicine. In addition to his research, he practices internal medicine and teaches medical ethics and the history of medicine.

IAN A. LI is Assistant Professor in the Department of the History of Medicine at Johns Hopkins University School of Medicine. Li is the author of *Body Maps: Improvising Meridians and Nerves in Global Chinese Medicine* (Johns Hopkins University Press, 2025). Li's work as a filmmaker and researcher explores the ontological and epistemic implications of practiced multiplicity within and across knowledge systems. Li is the founder of the Medicine, Race, Democracy Lab (mrdlab.org).

AMANDA L. MAHONEY is Director and Chief Curator of the Dittrick Medical History Center at Case Western Reserve University. As a historian of medical technology and nursing, her work explores the implementation and day-to-day use of clinical tools and the ethical, technical, and interpersonal challenges surrounding technologies in the United States during the twentieth century.

JESSICA MARTUCCI is the Curator of the Barbara Bates Center for the Study of the History of Nursing and an Adjunct Associate Professor in the History and Sociology of Science Department at the University of Pennsylvania. Her work examines how "outsider" groups experience and shape knowledge production, institutions, policies, and practices in American science and health care. Her first book is *Back to the Breast: Natural Motherhood and Breastfeeding in America* (University of Chicago Press, 2016).

RICHARD A. MCKAY is College Lecturer and Director of Studies in History and Philosophy of Science at Magdalene College, Cambridge. His research has been published in the *Bulletin of the History of Medicine, Nature, Medical Anthropology,* and *Newsweek.* His first book, *Patient Zero and the Making of the AIDS Epidemic* (University of Chicago Press, 2017), was named a *Choice Reviews*

Outstanding Academic Title and made into the award-winning documentary film *Killing Patient Zero*.

APARNA NAIR is Assistant Professor, Department of Health and Society, University of Toronto Scarborough, and the Center for Global Disability Studies. Her forthcoming book, *Fungible Bodies* (University of Illinois Press), examines the relationships between disability, race, and empire. She also works on the histories of public health (smallpox vaccination in particular); the histories of disability, technology, and race in the Global South; the representations of disability in popular and material culture; and animal histories.

AYAH NURIDDIN is Assistant Professor in the History of Medicine at Yale University. She received her PhD in the history of medicine at Johns Hopkins University in 2021. She is currently working on a book manuscript that examines Black engagements with eugenics and racial science across the twentieth century. Her work has been published in *Historical Studies in the Natural Sciences, Journal of the History of Medicine and Allied Sciences*, and *The Lancet*.

SHARRONA PEARL is Andrews Chair of Interdisciplinary Studies at the John V. Roach Honors College at Texas Christian University. Pearl's most recent book is *Do I Know You? From Face Blindness to Super Recognition* (Johns Hopkins University Press, 2023), which is the third in her face trilogy, following *Face/On: Face Transplants and the Ethics of the Other* (University of Chicago Press, 2017) and *About Faces: Physiognomy in Nineteenth-Century Britain* (Harvard University Press, 2010). She has also recently published *Mask* with Bloomsbury Academic (2024).

BEATRIZ PICHEL is Associate Professor at the Photographic History Research Centre, De Montfort University, United Kingdom. Her book *Picturing the Western Front* (Manchester University Press, 2021) examines how photographic practices shaped war experiences. She is the Principal Investigator of the Ethics of Medical Photography Network, and she is working on the history and ethics of medical photographic practices in nineteenth-century France.

AHMED RAGAB is Associate Professor in the Department of the History of Medicine at Johns Hopkins School of Medicine. Ragab is a historian, physician, and documentary filmmaker, as well as the founding director of the independent Center for Black, Brown, and Queer Studies. He has published three books: *The Medieval Islamic Hospital: Medicine, Religion and Charity* (Cambridge University Press, 2015); *Piety and Patienthood in Medieval Islam* (Routledge, 2018); and *Medicine and Religion in the Life of an Ottoman Sheikh* (Routledge, 2019).

MARCO ANTONIO RAMOS is Assistant Professor in the History of Medicine and Department of Psychiatry at Yale University. His research and teaching focus on the history of mental health in Latin America, with an emphasis on health activism. He is currently writing a book on radical mental health and terror in Cold War Argentina. His writing has appeared in clinical, academic, and popular journals, including the *American Historical Review*, the *Bulletin of the History of Medicine*, and the *New England Journal of Medicine*.

SUSAN M. REVERBY is the Marian Butler McLean Professor Emerita in the History of Ideas and Professor Emerita in Women's and Gender Studies, Wellesley College. A historian of American women, health care, nursing, medicine, and public health, she is the author of three books and editor or coeditor of five volumes, as well as numerous scholarly and public articles. Her current project is a dual biography of her paraplegic grandmother and New York's Montefiore Hospital in the mid-twentieth century.

JONATHAN SADOWSKY is Castele Professor of the History of Medicine and Associate Director of the Program in Medicine, Society, and Culture in the Department of Bioethics at Case Western Reserve University. He is the author of *Imperial Bedlam: Institutions of Madness and Colonialism in Southwest Nigeria* (University of California Press, 1999); *Electroconvulsive Therapy in America: The Anatomy of a Medical Controversy* (Routledge, 2016); and *The Empire of Depression: A New History* (Polity, 2023). He is coeditor of the six-volume *Cultural History of Madness* (Bloomsbury, forthcoming).

SHELLEY ANGELIE SAGGAR is a postdoctoral researcher at Goldsmiths, University of London. Her research examines cultural representations of the museum in comparative Indigenous literature and film. Shelley is the founder of *The Decolonial Dictionary* (https://decolonialdictionary.wordpress.com/), a resource that unpacks postcolonial theory in museum contexts. Alongside her academic research, Shelley regularly works with museums and cultural institutions on a range of curatorial and collections initiatives relating to colonial inheritances and community connections.

NICOLE LEE SCHROEDER is Assistant Professor of History at Kean University in Union, New Jersey. She writes, researches, and teaches in disability studies and early American history. Her research focuses on disability advocacy and evolving definitions of disability in early American history. Schroeder is the founder of the Disabled Academic Collective, a community with over 900 members that provides support to disabled academics at all career stages.

AISLING SHALVEY is a historian of medicine, specializing in pediatrics, pathology, and reproduction. She is a postdoctoral researcher and formerly worked with the Historical Commission for the Reichsuniversitat Strassburg, and the German National Academy of Science and Medicine at the Leopoldina. Her book, *Paediatrics in the Reichsuniversität Straßburg: Children's Medicine at a Bastion of Nazi Ideology,* was published in 2023 with Exeter University Press.

KYLIE M. SMITH is Associate Professor and Director of the Center for Healthcare History and Policy at Emory University in Atlanta, Georgia. She is the author of *Talking Therapy: Knowledge and Power in American Psychiatric Nursing* (Rutgers University Press, 2020). Her new project, *Jim Crow in the Asylum: Psychiatry and Civil Rights in the American South,* will be published by the University of North Carolina Press in 2025.

AMY C. SULLIVAN is a historian of drugs, medicine, race, gender, and childhood at Macalester College. Her research projects highlight narratives rooted in social change, community, and healing through oral history, public history, and digital archives. She is the author of *Opioid Reckoning: Love, Loss, and Redemption in the Rehab State* (University of Minnesota Press, 2021).

COURTNEY E. THOMPSON is Associate Professor in History at Mississippi State University. Her first book, *An Organ of Murder: Crime, Violence, and Phrenology in Nineteenth-Century America,* was published in 2021 with Rutgers University Press. Her work has been published in *Bulletin of the History of Medicine, Journal of Women's History, Journal of the History of Medicine and Allied*

Sciences, and *Social History of Medicine.* Her current research focuses on the intersection of the history of emotions and the history of medicine in nineteenth-century America.

LAUREN MACIVOR THOMPSON is a historian of the gender politics of late nineteenth- and early twentieth-century American medicine, law, and public health. She is Assistant Professor of History and Interdisciplinary Studies at Kennesaw State University and serves as the faculty fellow at the Georgia State University College of Law's Center for Law, Health and Society. Her book on the medical politics of the early birth control movement is forthcoming with Rutgers University Press.

SHANNON K. WITHYCOMBE is Associate Professor of History at the University of New Mexico. She is the author of *Lost: Miscarriage in Nineteenth-Century America* (Rutgers University Press, 2018) and has published on the history of miscarriage, pregnancy, and childbirth for academic and public audiences. She is currently researching the history of prenatal health care in the United States, the racist biology involved, and how to ethically amplify the work of Black women combatting the dangers of this public health program.

Figures are indicated by page numbers in italics.

AAHM. *See* American Association for the History of Medicine

ableism: academic, 7–8, 49, 51–55; in archives and museums, 117, 130; digital health humanities projects, 244, 251; electro-convulsive therapy and, 163; trigger warnings for, 269–70

abortion, 175

accessibility and accommodations, 48–58; academic ableism and, 52–55; advocacy for, 7, 54–55; archives and museums, 117; conference presentations, 49–51; confidentiality issues, 52–53; crip time and, 53; denials, effect on academic careers, 49–54; museum representations of disability and, 124; virtual attendance at conferences, 50–51

accountability: citation and, 223; to disabled communities, 52, 55; for ethical standards, 8, 146, 239; IRB reviews and, 144; peer reviews and, 235, 239; reparatory history and, 39–41, 211, 327

acknowledgments, 162, 201, 230–31n1, 238

activism. *See* advocacy and activism

acupuncture, 310–11

addict label in historical writing, 199–208; Lexington Narcotic Farm, 201–3; opioid epidemic and, 203–6; persistence of, 199–200; person-first language and, 200

advocacy and activism: for accommodations, 7, 54–55; by descendant communities, 91, 93n12, 94n19, 211; presentism and, 35; public health historians and, 31; queer history of nursing, 59–67

African Americans: Black Lives Matter movement, 1, 45, 223; Black power method, 152–58; centering in teaching History of Medicine, 257–65; citational justice and, 222–25; Cite Black Women Collective, 223, 229n1; hesitance to get vaccines, 258; National Black Nurses Association, 64–65; remains used in teaching and research, 87. *See also* racism and racial hierarchy; slavery and enslaved persons; Tuskegee Study of Untreated Syphilis in the Negro Male

Agamben, Giorgio, 142

agency: colonialism and, 74–75, 77–79, 81–82; commercialization of human remains and, 135; of museum staff, 120–21; of subjects of writing, 204

Ahmed, Sara, 225

A must-read book from **HOPKINS PRESS**

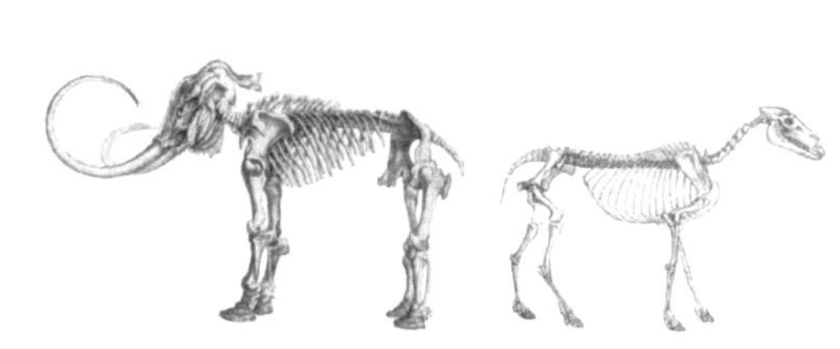

"Offers an imperative opportunity to reflect on the relationship between scientific knowledge and democratic life."

—Kathryn Lofton, Yale University

 JOHNS HOPKINS UNIVERSITY PRESS | PRESS.JHU.EDU |   

www.ingramcontent.com/pod-product-compliance
Ingram Content Group UK Ltd.
Pitfield, Milton Keynes, MK11 3LW, UK
UKHW041147190526
471241UK00002B/50

9 781421 452265